ESSENTIAL IDEAS IN
INORGANIC CHEMISTRY

By the same author

ESSENTIAL IDEAS IN ORGANIC CHEMISTRY
ESSENTIAL IDEAS IN PHYSICAL CHEMISTRY

ESSENTIAL IDEAS IN INORGANIC CHEMISTRY

D. E. WILSON, M.A., Ph.D.
Deputy Head, Maidstone Grammar School
Formerly Head of Chemistry, Manchester Grammar School

HODDER AND STOUGHTON
LONDON SYDNEY AUCKLAND TORONTO

British Library C.I.P.

Wilson D E
Essential ideas in inorganic chemistry.
1. Chemistry, Inorganic
I. Title
546 QD151.5

ISBN 0-340-20828-7

First printed 1978

Printed in Great Britain for
Hodder and Stoughton Educational,
a division of Hodder and Stoughton Ltd.,
Mill Road, Dunton Green, Sevenoaks, Kent
by Richard Clay (The Chaucer Press) Ltd,
Bungay, Suffolk

Preface

This book deals with inorganic chemistry at 'A' level in such a way as to take account of the revision of syllabuses carried out in recent years by most examining boards. Concepts such as those of the Born–Haber cycle are used to underpin the detailed comparison, group by group, of the elements. In this way the subject is presented as an integrated unit and not as a selection of individual facts.

Nevertheless, chemistry is a subject in which facts abound, and many find it difficult to grasp a concept except in terms of a concrete example. To include factual material in traditional descriptive style throughout, and include an extensive discussion of principles, would produce a work of considerable length and expense. Consequently compression has been achieved by the extensive use of flowsheets and tabulation of data. For those who, like the author, prefer to acquire and retain their knowledge by visual analogy, this is an additional advantage, over and above the easy information retrieval arising from the use of these techniques.

The aim of the book has been to examine the 'what' in order to stimulate the 'why' and the 'how'. For this reason, most of the questions are designed to encourage critical habits of thought. Candidates preparing for examination are better served by this type of question, coupled to practice from actual papers shortly before the day, than by a continual and monotonous diet of old papers throughout the course.

Finally, I would like to thank my wife, for putting up with a husband so often absent in thought if not in the flesh; to Isobel Willmott for typing, and especially for taking such pains with the unfamiliar words and symbols; and to Graham Hill, Senior Science Master at Bristol Grammar School, for detailed criticism and advice.

DAVID WILSON
1978

Contents

1
Introduction

1.1 THE SCOPE AND ARRANGEMENT OF THIS BOOK

This book is concerned with inorganic chemistry, that is, the study of the individual properties of elements and compounds, other than those containing both carbon and hydrogen. Because of the multiplicity of carbon–hydrogen compounds, their properties are best considered as a separate division of chemistry, called organic chemistry. The other major division, physical chemistry, is concerned with the more general properties common to all elements and compounds.

> Inorganic chemistry would be concerned with the properties of such compounds as NaOH, $CaCl_2$, SO_2 and CO_2.
> Organic chemistry would be concerned with the properties of such compounds as CH_3Cl, H_2CO, C_3H_8 and C_2H_6O.
> Physical chemistry would be concerned with the boiling points, solubilities or bonding of any of the above compounds.

These divisions should not be regarded as watertight boundaries.

The aim of this book is to present inorganic chemistry in terms of the physical principles by which the properties of inorganic compounds are classified and predicted. Each chapter is concerned with trends, classification, and reasons for them; these might be likened to the bone and sinew of a body. The flesh and outward appearance, in this analogy, are the facts and individual peculiarities of each reaction. Although less emphasis is placed on these, detailed summary tables appear at the end of each appropriate chapter. In most cases, one chapter covers the elements of one group of the periodic table.

As a result of this approach, a number of topics regarded as part of physical chemistry are referred to continually in the text and are used as unifying themes. For a full treatment of these you should consult a physical chemistry textbook. The following summary outlines the topics and the depth to which each is referred.

(1) *Electronic structure and bonding:* the nature of electrons, the structure of the atom including s, p and d as terms to describe energy levels, ionic, covalent and co-ordinate bonding, dipole and hydrogen-bond interaction.

(2) *Thermodynamic cycles:* the use of Hess' Law, the use of the Born–Haber cycle including the meaning of the terms in it, the meaning and use of bond enthalpies, the significance and limitation of enthalpy change as a guide to reaction.

(3) *Redox potentials:* the signficance and limitation of redox potential as a guide to reactions.

(4) *Acid–base theory:* interpretation of acid–base reactions in terms of proton transfer, the nature of $H^+(aq)$ water, as a source of H^+ or OH^- ion, pK_a as a method of comparing acid strength. A fifth unifying principle, the periodic classification, is dealt with at length in chapter 2.

Where industrial chemistry is described, some comment is made about the economic factors involved. Remember that the chemical industry operates in order to make profits, and not to illustrate the finer points of chemistry on the tonnage scale. The reactions and purifications needed to make a particular product are often chosen differently in industry and in the laboratory. The differences are best studied by means of some of the excellent case studies now available.

The case study approach is also recommended for a study of pollution. Problems of pollution are often simply solved by the application of technology (e.g. the removal of sulphur dioxide from flue gases by an alkali wash) but invariably add to the cost of the product or activity. The really difficult problem is economic: what are you prepared to pay or do to have the nuisance removed?

1.2 THE NAMES OF INORGANIC COMPOUNDS

In most cases, the oxidation state of the central element is used to name a compound. The oxidation state is a measure of the degree of oxidation of an element, since it shows how many electrons have been added to or removed from it. To find the oxidation state of an atom in a molecule or ion, it is assumed (quite fictitiously!) that the molecule is ionic. The most electropositive elements (H, alkali metals) or most electronegative elements (O, F, Cl, etc.) are given the charge that they would normally carry. The atom under examination is then given the charge necessary to account for the actual charge on the molecule (zero) or ion. Roman numerals are used to indicate oxidation states, to prevent confusion with real charge.

The following examples show how this works in practice:

(i) CrO_3 Take O as O^{2-}; $3O^{2-}$ make a total of 6—.
Net charge on molecule is zero.
Therefore the oxidation state of chromium must be 6+, written Cr^{VI}.

(ii) HIO_4 Take H as H^+ and O as O^{2-}.
One H^+ and $4O^{2-}$ make a total of 7—.
Net charge on molecule is zero.
Therefore the oxidation state of iodine must be 7+, written I^{VII}.

(iii) $CuCl_4^{2-}$ Take Cl as Cl^-; $4Cl^-$ make a total of 4−.
Net charge on ion is 2−.
Therefore the oxidation state of copper must be 2+, written Cu^{II}.

(iv) VO_2^+ Take O as O^{2-}; $2O^{2-}$ make a total of 4−.
Net charge on ion is 1+.
Therefore the oxidation state of vanadium is 5+, written V^V.

(v) NH_2OH Take H as H^+ and O as O^{2-}.
$3H^+$ and one O^{2-} make 1+.
Net charge on molecule is zero.
Therefore the oxidation state of nitrogen is 1−, written N^{-I}.

The oxidation state may be positive, negative, or zero as is the case for any uncombined element.

Since oxidation corresponds to losing electrons, i.e. gaining positive charge, an increase in oxidation state corresponds to oxidation. Conversely, a decrease in oxidation state corresponds to reduction.

The rules for naming compounds are as follows:

(a) *Substances which exist as individual covalent molecules.* Oxidation states are not used. 'Di', 'tri', 'tetra', 'penta' and 'hexa' are used to show 2, 3, 4, 5 and 6 substituents on an atom.

$SiCl_4$	silicon tetrachloride
SF_6	sulphur hexafluoride
NO	nitrogen oxide
NO_2	nitrogen dioxide

(b) *Substances which exist as giant lattices, whether covalent or ionic.* The oxidation state is written in brackets after the appropriate element

CrO_3	chromium(VI) oxide
SiO_2	silicon(IV) oxide

(c) *Substances which contain ions.* The name of the ion is built up sequentially. The name of the other ion present is then added, e.g. K_2CuCl_4, potassium tetrachlorocuprate(II). Here are some other examples

$Cu(H_2O)_6^{2+}$ SO_4^{2-}	hexaquacopper(II) sulphate(VI)
HIO_4	iodic(VII) acid
$K_4Fe(CN)_6$	potassium hexacyanoferrate(II) (note II for oxidation state, not IV for 4−)

The definition of oxidation as loss of electrons can be applied easily if the oxidation state is known. For example:

Zn(0) $\longrightarrow$ Zn(II)Cl_2 : zinc goes from 'Zn^0' to 'Zn^{2+}' losing 2 electrons: oxidised
$H_2S(-II) \longrightarrow$ S(0) : sulphur goes from 'S^{2-}' to 'S^0' losing 2 electrons: oxidised

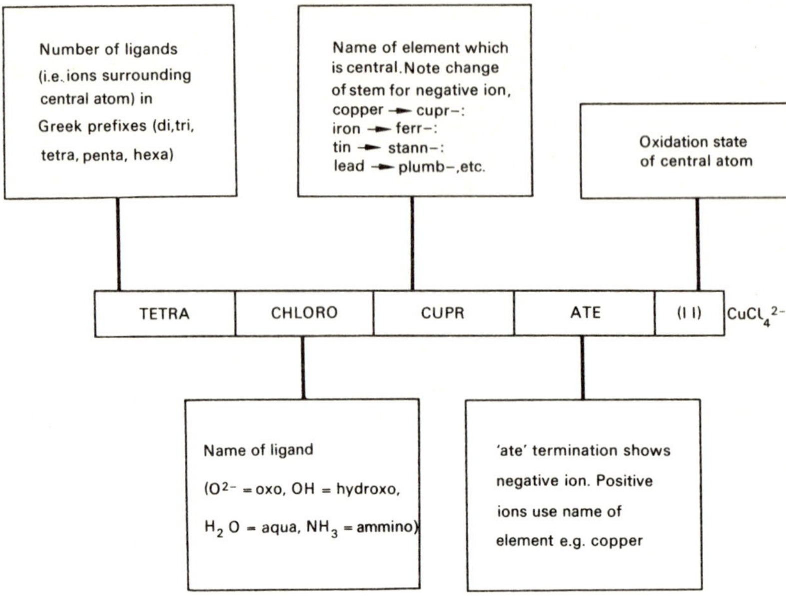

More complex examples:

$MnO_4^- \longrightarrow MnO_2$: manganese goes from 'Mn^{7+}' to 'Mn^{4+}' gaining 3 electrons: reduced

$NO_3^- \longrightarrow NH_3$: nitrogen goes from 'N^{5+}' to 'N^{3-}' gaining 8 electrons: reduced

1.3 BALANCING EQUATIONS

Since electrons are never freed in chemical reactions, but transferred entirely from reductant to oxidant, the stoichiometry of redox reactions can be calculated instantly. Taking an example from above, manganate(VII) can be used to oxidise hydrogen sulphide.

If MnO_4^- gains 3 electrons in becoming MnO_2
and H_2S loses 2 electrons in becoming S

then, without further ado, $2MnO_4^-$ reacts with $3H_2S$ (written $2MnO_4^- \equiv 3H_2S$). This corresponds to a transfer of a total of six electrons. To take another example, nitrate(V) can be reduced to ammonia by aluminium powder:

$$NO_3^- \longrightarrow NH_3 \quad ; \text{ oxidation state change } +\text{V} \longrightarrow -\text{III} = -8$$
$$Al \longrightarrow Al(OH)_6^{3-} \; ; \text{ oxidation state change } 0 \longrightarrow +\text{III} = +3$$

therefore 8 Al reacts with $3NO_3^-$ (written $8Al \equiv 3NO_3^-$). This corresponds to a transfer of twenty-four electrons.

This technique only balances reactions: it does not tell you the conditions for a particular reaction (e.g. nitrate(V) and aluminium need to be in sodium hydroxide solution). Nor does the technique tell you what product is formed (e.g. manganate(VII) is only reduced to manganese(IV) oxide in neutral solution but in acid solution $Mn(H_2O)_6^{2+}$ is formed).

For many purposes the above equivalence is sufficient. In other cases a full balanced equation is necessary. Sometimes this may be achieved by inspection:

$$H_2S^{-II} \longrightarrow S^0 \text{ and } S^{IV}O_2 \longrightarrow S^0$$
$$\therefore 2H_2S \equiv SO_2 \text{ or } 2H_2S + SO_2 \longrightarrow 3S + 2H_2O$$

More often, and especially for reactions in aqueous solution, 'half-equations' have to be built up. The following example shows how this can be done. Manganate(VII) ions can be reduced to hydrated manganese(II) ions in acid solution by iron(II) ions.

(1) Establish the change in oxidation state.	$MnO_4^-(VII) \longrightarrow Mn^{2+}(II)$
(2) Enter in the number of electrons.	$MnO_4^- + 5e^- \longrightarrow Mn^{2+}$
(3) If in acid, add enough protons to balance change (in alkali, hydroxide ions must be used).	$MnO_4^- + 8H^+ + 5e^- \longrightarrow Mn^{2+}$
(4) The half equation will now balance if the appropriate number of molecules of water are entered.	$MnO_4^- + 8H^+ + 5e^- \longrightarrow Mn^{2+} + 4H_2O$
The other half equation is simpler:	$Fe^{2+} \longrightarrow Fe^{3+} + e^-$
(5) Add 5 times the second half equation to the first, so that the number of electrons transferred in each equation is the same.	$MnO_4^- + 8H^+ + 5e^- + 5Fe^{2+} \longrightarrow Mn^{2+} + 4H_2O + 5Fe^{3+} + 5e^-$
(6) Cancelling the electrons gives a balanced equation.	$MnO_4^- + 8H^+ + 5Fe^{2+} \longrightarrow Mn^{2+} + 4H_2O + 5Fe^{3+}$

If potassium manganate(VII), sulphuric(VI) acid, and iron(II) sulphate(VI) are used, the full equation can now be written following the pattern of the last equation

$$KMnO_4 + 4H_2SO_4 + 5FeSO_4 \longrightarrow MnSO_4 + 4H_2O + \tfrac{5}{2}Fe_2(SO_4)_3 + \tfrac{1}{2}K_2SO_4$$

Here is another example. Potassium dichromate(VI) in dilute acid oxidises hydrogen sulphide to sulphur.

(i) $Cr_2O_7^{2-}$(VI) $\longrightarrow 2Cr^{3+}$(III) $\qquad$ $H_2S(-II) \longrightarrow S(0)$

(ii) $Cr_2O_7^{2-} + 6e^- \longrightarrow 2Cr^{3+}$ $\qquad$ $H_2S \longrightarrow S + 2e^-$

(Note: $2 \times 3e^-$ required since 2 moles of Cr change by 3 oxidation states)

(iii) $Cr_2O_7^{2-} + 6e^- + 14H^+ \longrightarrow 2Cr^{3+}$ $\qquad$ $H_2S \longrightarrow S + 2e^- + 2H^+$

(iv) $Cr_2O_7^{2-} + 6e^- + 14H^+ \longrightarrow 2Cr^{3+} + 7H_2O$ $\qquad$ $H_2S \longrightarrow S + 2e^- + 2H^+$

(v) $Cr_2O_7^{2-} + 14H^+ + 6e^- + 3H_2S \longrightarrow 2Cr^{3+} + 7H_2O + 3S + 6e^- + 6H^+$

(vi) $Cr_2O_7^{2-} + 14H^+ + 3H_2S \longrightarrow 7H_2O + 3S + 6H^+$

2
The Periodic Table

2.1 INTRODUCTION

The atomic number of an element is the number of protons in the nucleus. If elements are arranged in order of increasing atomic number, it is found that elements having similar properties recur at regular intervals, at first every eighth element, then every eighteenth, and finally every thirty-second. For example

$$_{3}Li — {}_{11}Na — {}_{19}K — {}_{37}Rb — {}_{55}Cr$$
or $$_{4}Be — {}_{12}Mg — {}_{20}Ca — {}_{38}Sr — {}_{56}Ba$$
or $$_{9}F — {}_{17}Cl — {}_{35}Br — {}_{53}I$$

This recurrence of properties is now built into the familiar periodic table (end-papers) and seems such an obvious way to classify the elements that it is perhaps worth recalling the difficulties of Mendeleyev, who first proposed the scheme.

The information that he had to work with was:

(1) a knowledge of the properties of a number of elements—but this was incomplete. Germanium and gallium were two elements of relatively low atomic mass still to be discovered.

(2) a knowledge of the atomic masses of these elements—but several of these were wrong. Some were wrong because a wrong valency had been assumed for the element, as in the case of beryllium. Beryllium was known to form an oxide containing 36 per cent beryllium and thought to be Be_2O_3. This is consistent with a relative atomic mass of 13.5 $\left(\frac{(2 \times 13.5)}{(2 \times 13.5) + (3 \times 16)} \times 100 = 36\right)$; in fact the formula is BeO and the atomic mass is 9. If the former mass applied, beryllium would be placed between carbon and nitrogen; if the latter, then it would be placed between lithium and boron. Other atomic masses were wrong because of analytical difficulties, e.g. osmium was thought to have an atomic mass of 198.6; it was checked in 1912 and found to be 190.9.

(3) a knowledge of the periodic variation of molar volume, set out by Lothar Meyer a few years before—but this property shows less clear-cut trends than atomic radius, discussed below. Atomic radius is a function only of the electronic structure of the element: but the volume of one mole is complicated by the type of bonding in the element and/or the packing of atoms into a crystal.

Mendeleyev's bold proposals, involving the alteration of established atomic masses and the leaving of gaps for new elements, got a hostile reception. However, the accuracy of his predictions (table 2.1) eventually won support. One prediction not borne out concerned the pair of elements tellurium and iodine. In this case the atomic mass of the first element is greater than the second, although its position in the periodic table is confirmed by its properties. Mendeleyev thought that the atomic mass was wrong. Today it is recognised that isotopes of tellurium contain rather more neutrons than isotopes of iodine. This affects the atomic mass, but not the atomic number used for ordering elements today. Argon/potassium and cobalt/nickel are another two pairs of elements showing the same behaviour.

Property	*Atomic mass*	*Density*/ gcm^{-3}	*Valency*	*Boiling point* XCl_4/°C	*Boiling point* $X(C_2H_5)_4$/°C
Eka-silicon, predicted	72	5.5	4	90	160
Germanium found	72.6	5.47	4	86	163.5

Table 2.1 Mendeleyev's predictions for germanium.

2.2 PROPERTIES SHOWING PERIODIC BEHAVIOUR

2.2.1 **Group behaviour**

Elements falling in the same vertical column are known as a group and show considerable similarity of behaviour. Where there are differences, there is commonly a gradual trend from the element at the top of the column to the one at the bottom. It was on this basis that Mendeleyev made his predictions. The behaviour of groups is considered in the subsequent chapters of this book, and the remainder of this chapter is devoted to variations across the periodic table.

2.2.2 **Electronic structure**

Because the elements in the periodic table are placed in order of increasing atomic number (reading left to right and top to bottom) each atom has one more proton in its nucleus than the previous one. Hence the number of electrons must increase in the same way in order to maintain electrical neutrality. The electrons form the outer regions of the atom, and so are responsible for interactions with other atoms, namely, the chemical properties.

Each electron falls into a vacancy in one of the energy levels possessed by the atom. When one level is full, the next electron must go into a level further from the nucleus, where it is less firmly held. The energy levels may take up to two electrons (s-type levels), six electrons (p-type levels), or ten electrons (d-type levels), and so on. The s-, p- and d-type levels are characterised by the symmetry of the regions of space, or orbitals, in which that type of electron is to be found. s orbitals are spherically symmetrical, p orbitals are cylindrically symmetrical and so down to lower orders of symmetry. The first shell nearest the nucleus contains two electons, both in an s orbital written s^2; the second shell contains s and p electrons making a total of eight; the third shell contains s, p and d electrons to a total of eighteen, etc., so that the nth shell contains $2n^2$ electrons. With this information the shape of the periodic table can be explained.

The notation employed is

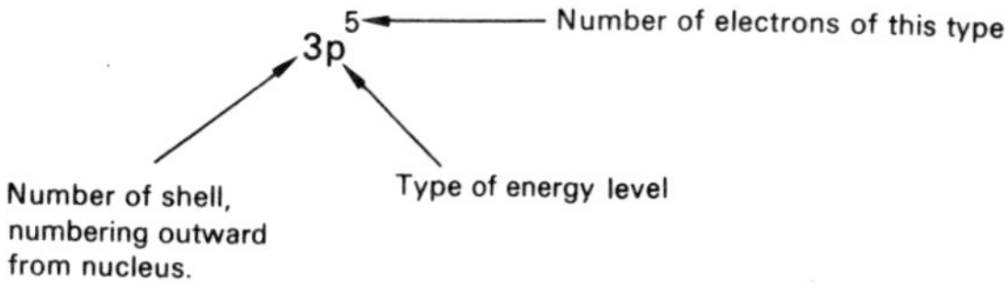

Period	*Comment*
1	The two elements H and He contain one and two s electrons respectively. This completes the first shell
2	The elements Li to Ne are built up by filling the second shell with electrons. The eight elements reflect eight electrons ($s^2 + p^6$)
3	Essentially the same as period 2
4	The third shell awaits completion while two s electrons of the fourth shell are added (K, Ca); then the third shell is completed with ten d electrons (ten elements Sc–Zn) before continuing the fourth shell with six p electrons (six elements Ga–Kr)
5	Essentially the same as period 4

Table 2.2 Periods and electronic structure.

Perhaps the key point to emerge from the above is that the noble gases (He–Rn) act as a reference stage. Their electronic structure in each case (except He) is the 'stable' octet of two s and six p electrons. (Helium cannot have p electrons, since only the first shell is in use, but it has nevertheless a stable full shell of two s electrons.)

The three major blocks which appear on inspecting the periodic table are related to the noble gas structure as follows:

(a) the first and second columns on the left (groups I and II). These elements contain one and two electrons, respectively, more than the noble gas 'core'. These extra electrons are in an s orbital and the elements are said to be s-block elements.

(b) the six columns on the left (groups III–VIII). Here the p orbitals are being filled, and the elements contain between one p electron (group III) and six p electrons (group VIII). These are the p-block elements.

(c) the elements in the centre. Although possessing an exterior similar to that of an s-block element, each element differs from the previous one in having one extra d electron. Thus Sc is made up of an Ar core, plus the two s electrons needed to create Ca, plus one d electron

$$Sc \equiv [Ar] + 4s^2 + 3d^1$$
$$Ti \equiv [Ar] + 4s^2 + 3d^2 \quad \text{and so on.}$$

A full list of electronic structures is given in appendix 2. A key point, which must be emphasised, is the strict similarity between elements in any one group. An electronic structure is common to a group, for example

$$Li = [He] + 2s^1,\ Na = [Ne] + 3s^1,\ K = [Ar] + 4s^1,\ Rb = [Kr] + 5s^1$$
or $C = [He] + 2s^2 2p^2$, $Si = [Ne] + 3s^2 3p^2$, $Ge = [Ar] + 4s^2 4p^2$, $Sn = [Kr] + 5s^2 5p^2$

2.2.3 **Bonding**

The types of bonds formed by an element depends closely on its electronic structure, and hence are related to the periodic table.

Electron arrangement	*Reasoning*	*Example*
s and p orbitals exactly full: eight electrons in outer shell (two electrons for He in first shell)	An extra electron has to go into a higher shell and would be weakly held: removal of electrons is also unlikely (see ionisation energy in next section). Therefore inert	Noble gases (see below in table) (Chapter 9)
One or two more electrons than the 'stable' noble gas core	The extra electron has gone into a shell further out than any previous, with only one unit increase in the nuclear charge. It is weakly held and readily lost to form cations X^+ or X^{2+}	Alkali and alkaline earth metals (Chapter 3)
One or two electrons less than the 'stable' noble gas core	There is a vacancy which available electrons can easily occupy. Hence reactive, forming anions X^- or X^{2-}	Oxygen, sulphur and halogens (Chapters 7 and 8)
About 4 electrons in outer shell, i.e. both half full and half empty	Removal of four electrons requires more energy than can be regained in a chemical reaction. No ions, only covalent bonds formed	Group 4 (Chapter 5)

Electron arrangement	*Reasoning*	*Example*
Electrons of highest energy in a 'd' orbital	The 3d orbital occupies a position in space closer to the nucleus than the 4s shell, which is already occupied. Electrons are removed easily because of crowding, but subject to the criterion that removing the fourth or later electrons from a highly charge positive ion is too expensive in energy terms. No definite number of electrons removed, therefore variable valency	Transition metals (Chapter 1)

Table 2.3 Possibilities of bonding.

These considerations also allow one to predict the principal valency of an element. For ionic compounds, the ion is usually formed by gaining or losing the number of electrons needed to establish the noble gas octet. By inspection, this creates a valency equal either to the number of the group, or to the difference between this number and eight. Covalent compounds may also display these two valencies, either by sharing all their electrons (in which case the valency is the number of the group) or by sharing just enough to make up a full shell (in which case the valency becomes the difference above). For instance, consider the oxides and hydrides in table 2.4.

Group	*Number of electrons in outer shell*	*Principal valencies*	*Oxide*	*Hydride*
I	1	1	Na_2O	NaH
II	2	2	MgO	MgH_2
III	3	3	Al_2O_3	not known
IV	4	4	SiO_2	SiH_4
V	5	3, 5	P_2O_5	PH_3
VI	6	2, 6	SO_3	SH_2
VII	7	1, 7	Cl_2O_7	ClH
VIII	8	0	not formed	not formed

Table 2.4 Principal valencies.

2.2.4 **Atomic radius**

Consider sodium in comparison with the previous element, neon. The outermost electron in sodium has to go in a shell further out than the previous electrons, which screen it from the attraction of the nucleus. It is therefore weakly held, and has a wider orbit, thus increasing the atomic radius.

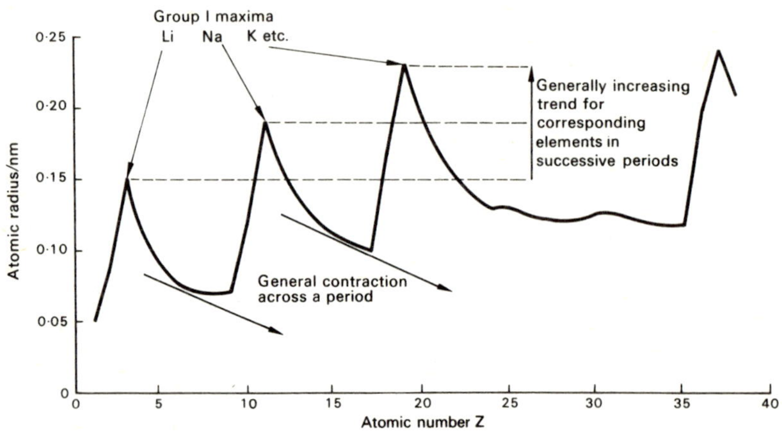

Fig. 2.1 Atomic radius.

As the period is traversed from sodium to argon, the extra protons acting on electrons in the same shell decrease the radius steadily, as shown by the figures in the table, and by the graph. There is a periodic variation, with a maximum recurring in group I.

Period 2	Li	Be	B	C	N	O	F	Ne
Atomic radius/nm	0.15	0.11	0.09	0.077	0.074	0.074	0.072	0.12
Period 3	Na	Mg	Al	Si	P	S	Cl	Ar
Atomic radius/nm	0.19	0.16	0.13	0.12	0.11	0.104	0.100	0.17

Table 2.5 Atomic radii of some elements (single atom value used if available, otherwise value for covalently bonded atom).

2.2.5 **Ionisation energy**

The ionisation energy is the energy required for the process $X(g) \longrightarrow X(g)^+ + e^-$. It is one of the major factors in determining whether or not an element forms cations (for a full discussion see chapter 2). The outermost electron of a group I element, the first in a period, is easily removed compared to that of previous elements because it is so much further from the nucleus. The electrostatic attraction between the positive nucleus and the negative electron decreases sharply with distance. Although the positive charge on the nucleus is larger, compared to elements in the previous group, the increased attraction is not sufficient to counteract the effect of the increased radius.

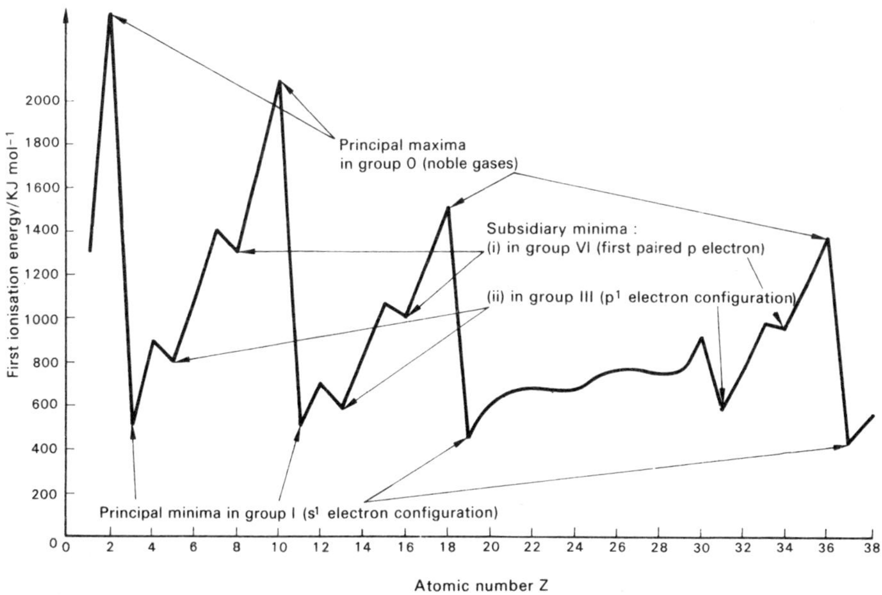

Fig. 2.2 First ionisation energy.

Successive elements in a period, compared to the first, show larger ionisation energies. This is related to the decrease in atomic radius discussed above, since the comparative closeness of the nucleus and the electron results in a greater attraction. The graph shows a general periodic trend with a minimum in group I.

Superimposed on this general trend are two minor effects. After the s-energy level is filled, the next electron has to go into the p level, which is of slightly higher energy. Less energy is therefore required to remove this first p electron. A secondary minimum appears at the third element in each period, after the two vacancies of the s orbital have been filled.

Another secondary minimum appears because of the 'pairing' effect. The six p electrons are arranged by pairs in three orbitals. When the first three electrons have been placed, one in each orbital, the fourth has to share an orbital with one of those already present. The repulsion ensuing allows this electron to be removed with less difficulty, so that group VI elements have a slightly lower ionisation energy than those on either side of them.

As might be expected, second and later ionisation energies are larger than the first, because of the extra work required to separate a negative electron from an already positive ion.

First ionisation energy	$X(g) \longrightarrow X^+(g) + e^-$
Second ionisation energy	$X^+(g) \longrightarrow X^{2+}(g) + e^-$
Third ionisation energy	$X^{2+}(g)^- \longrightarrow X^{3+}(g) + e^-$, etc.

Furthermore, there is a dramatic increase when an electron has to be removed from the next inmost shell, which would correspond to the second ionisation energy for sodium, the third for magnesium and so on. This sets an upper limit to the number of electrons which may be lost in a chemical reaction. The amount of energy required to cause ionisation of an electron from an inner shell cannot be regained from other reagents. In the case of the transition metals, in the centre of the table, there is no such definite limit because of the large number of d electrons which would have to be removed before reaching the next shell. Variable valency is then the rule.

Element	Ne	Na	Mg	Al	Si	P	S	Cl	Ar	K
1st ionisation energy	2080	494	736	577*	786	1060	1000†	1260	1520	420
2nd ionisation energy	3950	4560	1450	1820	1580	1900	2260	2300	2660	3070
Ionisation energy for removal of first inner shell electron	–	4560	7740							

* decreases for the removal of the first electron in the 3p orbital.
† decreases for the removal of the first paired p electron.

Table 2.6 Ionisation energy of second period elements. (*Values in* $kJmol^{-1}$.)

2.2.6 **Electron affinity**

Electron affinity is the energy change for the process $X(g) + e^- \longrightarrow X^-(g)$. It is a factor in determining whether an element forms anions in compounds. It is larger for elements on the right of the period and shows a maximum in group

VII. Here the incoming electron can be attracted closer to a small atom by a larger positive charge on the nucleus (see table 1.5). Second electron affinities are less exothermic than the first and may even be endothermic. The second incoming electron is repelled by the existing negative charge.

Element	Na	Mg	Al	Si	P	S	Cl	Ar
Electron affinity/ kJmol^{-1}	−53	−31	−44	−120	−74	−200	−349	+35

Table 2.7 Electron affinities of second period elements.

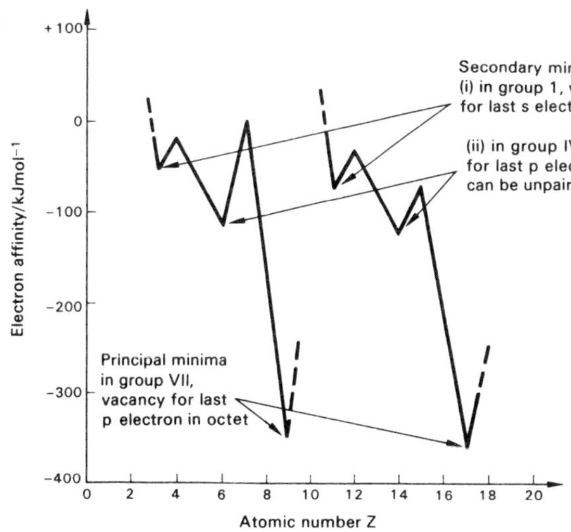

Fig. 2.3 First electron affinity (noble gases not plotted).

2.2.7 Melting points

Melting and boiling points, and heats of vaporisation, show a maximum in the centre of the table. These properties are concerned with the breaking of bonds between atoms, and vary with both the structure and bonding present. Elements to the left of a period are metals, with a close-packed structure (see chapters 3 and 10), and the many bonds which need breaking require large amounts of energy. Elements in the centre such as silicon form giant covalent lattices using several single bonds, which again need much energy for breaking. Elements to the right of the table form covalent bonds to one or more atoms. The resulting molecules are only weakly attached to each other, and are easily separated, having low boiling points and heats of vaporisation.

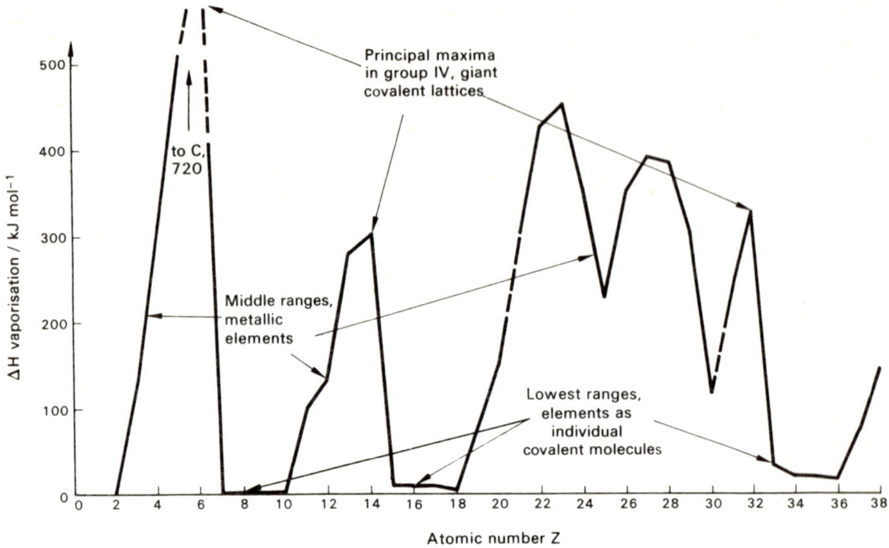

Fig. 2.4 Enthalpy of vaporisation (melting and boiling points show similar trends).

Element	*Structure of element*	*Bonding*	*Heat of vaporisation/ kJ mol^{-1} of atoms*
Na	Cubic lattice (8 neighbours)	Metallic	100
Mg	Close packed lattice (12 neighbours)	Metallic	130
Al	Close packed lattice (12 neighbours)	Metallic	280
Si	Giant molecule (4 neighbours)	Covalent	300
P	Giant covalent molecule in layers (3 neighbours)	Covalent	10*
S	Simple covalent molecule: a ring of atoms (2 neighbours)	Covalent within each molecule: molecules weakly attached to one another	10
Cl	Simple covalent diatomic molecule		10
Ar	monatomic		6

* vaporises as P_4 molecules: i.e. $\Delta H_{vap} = 40$ kJ for each mole of P_4.

Table 2.8 Heats of vaporisation of second row elements.

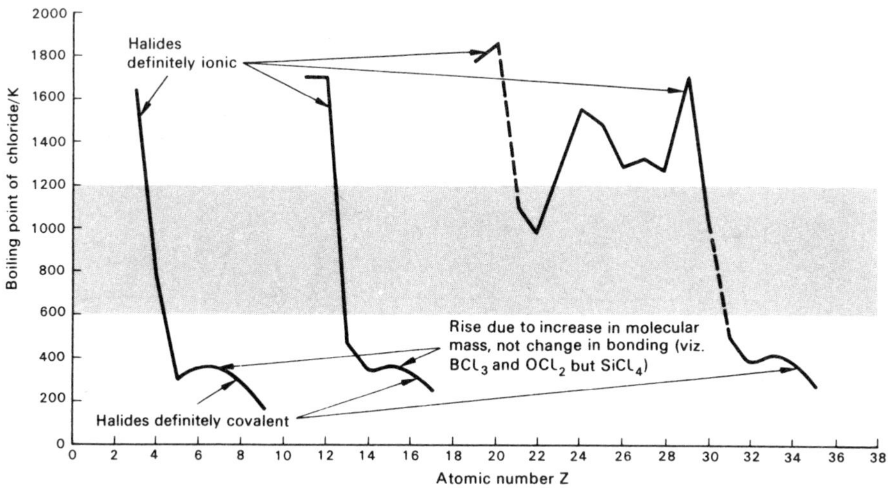

Fig. 2.5 Boiling point of chlorides (XCl_n, where n = number of group, except for transition metals where n = 3 (Sc, Ti, Cr), n = 2 (Fe ⟶ Zn). VCl_3 decomposes).

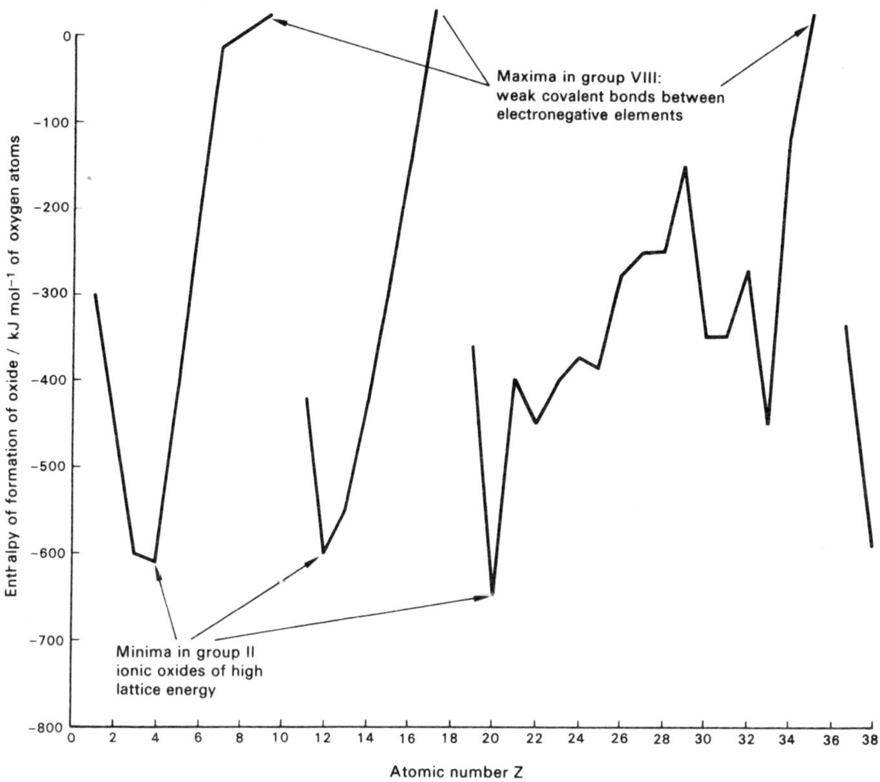

Fig. 2.6 Enthalpy of formation of most stable oxide.

METAL | NON-METAL

Property	*Typical metal will be*	*Na*	*Mg*	*Al*	*Si*	*P*	*S*	*Cl*	*Ar*
Ductility	Very ductile	Can be formed into wire in a hand press	Ribbon easily bent to shape	Bent easily with tools	Shatters on bending	Fractures on knife cut	Powders on crushing	Gas	Gas
Electrical conductivity	A good conductor ($\kappa > 0.1 \mu\Omega^{-1}\ cm^{-1}$	0.22	0.22	0.38	0.10	10^{-17}	10^{-23}	Negligible	Negligible
Formation of cations	Found as M^{n+}	All compounds cationic	A few covalent compounds, e.g. $Mg(CH_3)_2$	Only in AlF_3 and $Al(H_2O)_6^{3+}$	No cations	No cations	No cations	No cations	No cations
Reaction of hydrides with water	Alkaline in solution	Both react to form hydroxides $H^- + H_2O \longrightarrow H_2 + OH^-$		No reaction	No reaction	No reaction	Weakly acid	Strongly acid	–
Reaction of oxides with water	Alkaline in solution	Violent reaction to form NaOH (pH 14)	Reacts slowly to form $Mg(OH)_2$ (pH 11)	No reaction with water but will dissolve in acid or base	No reaction with water but will dissolve in strong alkali	Vigorous reaction to form strong acid	Vigorous reaction to form strong acid	Strongly acid	–
Bonding in chlorides	Ionic: chloride has boiling point >200°C	NaCl b. pt. 1740°C solid	$MgCl_2$ b. pt. 1680°C solid	$AlCl_3$ sublimes 180°C solid	$SiCl_4$ b. pt. 60°C liquid	PCl_3 b. pt. 75°C liquid	SCl_2 b. pt. 60°C liquid	Cl_2 b. pt. −35°C gas	–

METAL | NON-METAL

Table 2.9 Metals and non-metals.

2.2.8 **Metals and non-metals**

This classification divides elements into two rather ill-defined groups. Typical metallic properties are given in the table opposite. For the relation of these to electronic structure see chapter 11. Aluminium is considered to be a metal, and silicon a non-metal, but each has properties of the other group. The boundary between metals and non-metals lies across the periodic table roughly from boron to tellurium.

2.2.9

Other properties also show a periodic variation; in fact any chemical property picked at random is more likely to be related to the periodic table than not. As an example, enthalpies of formation of oxides are plotted on the graph (fig. 2.6).

QUESTIONS

1 Write out electronic structures in the form $1s^2$, $2s^2$, $2p^6$ . . ., etc. for elements of the following atomic numbers: 6, 9, 12, 18, 23, 34, 37. In each case show with reasons what type of bonds are most probably formed.

2 You are Mendeleyev. Write a paper to be delivered to the learned scientists of the day introducing your new periodic classification of the elements. Remember that many of them will not only be hostile to the idea, not having thought of it first, but also annoyed by your 'arbitrary' alteration of some atomic masses which they have themselves determined.

3 In each of the following cases, predict as much as possible about the chemistry of the first element given from the known properties of the others.

(a) ${}_{85}At$ from ${}_{17}Cl$, ${}_{35}Br$, ${}_{53}I$. (chapter 8)
(b) ${}_{31}Ga$ from ${}_{5}B$, ${}_{13}Al$. (chapter 4)
(c) ${}_{38}Sr$ from ${}_{20}Ca$, ${}_{56}Ba$. (chapters 3, 12)
(d) ${}_{27}Co$ from ${}_{26}Fe$, ${}_{28}Ni$. (chapter 11)

A data book will also be useful for this question.

4 State what periodic variation there is in each of the following properties, and *sketch* a graph to illustrate your statement. Show how this variation arises from our present knowledge of electronic structure.

(a) n in the formula XCl_n
(b) atomic volume
(c) the first ionisation energy $X(g) \longrightarrow X^+(g) + e^-$
(d) the second ionisation energy $X^+(g) \longrightarrow X^{2+}(g) + e^-$
(e) boiling point
(f) hardness
(g) physical state of oxide
(h) pH of oxide solution in water.

5 Calculate the oxidation state of the element *other than* H, Na, O, F or Cl in the following ions and compounds. Hence name the ion or compound, in the latter case consulting a data book in order to find out whether it exists as giant molecules or as an individual lattice.

Na_2CrO_4, SO_2, XeF_4, $Be(H_2O)_4^{2+}$, $Al(OH)_6^{3-}$, FeF_6^{3-},
PO_4^{3-}, AlH_4^-, TiO^{2+}, H_2SO_4, MnF_6^{4-}, IF_7, Na_2SO_3.

6 Balance the following equations by the method of sections 1.5 or 1.6. All reactants and products are given

(a) $Na_2CO_3 + H_3PO_4 \longrightarrow Na_3PO_4 + H_2O + CO_2$
(b) $CrO_4^{2-} + H^+ + I^- \longrightarrow Cr^{3+} + I_2 + H_2O$
(c) $V^{2+} + MnO_4^- + H^+ \longrightarrow VO_2^+ + Mn^{2+} + H_2O$
(d) $Ce^{3+} + BiO_3^- + H^+ \longrightarrow Ce^{4+} + Bi^{3+} + H_2O$
(e) $ClO_3^- + Fe^{2+} + H^+ \longrightarrow Cl^- + Fe^{3+} + H_2O$
(f) $MnO_4^- + H_2O_2 + H^+ \longrightarrow Mn^{2+} + O_2 + H_2O$
(g) $SO_2 + Fe^{3+} + H_2O \longrightarrow Fe^{2+} + SO_4^{2-} + H^+$
(h) $NO_3^- + Al + OH^- + H_2O \longrightarrow NH_3 + Al(OH)_6^{3-}$
(i) $(NH_4)_3PO_4 + Ca(OH)_2 \longrightarrow Ca_3(PO_4)_2 + NH_3 + H_2O$
(j) $Mg_3N_2 + H_2O \longrightarrow Mg(OH)_2 + NH_3$.

7 The following statements are not correct in all respects. There may be exceptions, ambiguities, omissions or deliberate mistakes. Write out a correct version of the statements, and comment on the theoretical background to it. A data book will be useful.

(a) Elements whose electronic configuration differs from that of a noble gas by one or two electrons form compounds which are always ionic.
(b) The first ionisation energy increases steadily from Li to Ne.
(c) The transition metals are those elements which have electrons in d orbitals.
(d) Molar volume decreases steadily from Na to Ar.
(e) Metal oxides form basic solutions in water.
(f) Non-metal hydrides are gases.
(g) Compounds which have the formula XO_2 are called X(IV) oxides.
(h) Only metals have positive oxidation states.
(i) The cation in a dilute acid solution is H_3O^+.
(j) A full electron shell contains eight electrons.

8 Arrange substances according to the property stated. Do not use published data, but justify your answer by reference to physical principles.

(a) B, Be, Li, Na, Ne according to first ionisation energy.
(b) $Al(OH)_3$, $C(OH)_4 \equiv [H_2CO_3.H_2O]$, $Ca(OH)_2$, KOH, $Mg(OH)_2$ according to pH of saturated aqueous solution.
(c) $AlCl_3$, HCl, KCl, NaCl, PCl_3 according to boiling points.
(d) Ar, C, Cl, Cs, Mg according to melting points.
(e) Li, Pb, Rb, Xe, Zn according to number of electrons short of noble gas structure.

3
The s-block elements

Two groups of cations

3.1 INTRODUCTION

The s-block elements are those of the first two groups of the periodic table, comprising in group I lithium, sodium, potassium, rubidium and caesium; and in group II, beryllium, magnesium, calcium, strontium and barium. In all cases the outermost electrons occupy a spherically symmetrical s orbital, one electron in the case of group I, and two electrons in the case of group II. Because these electrons are situated further out than all previous electrons in the atom, which form a noble gas core, they are only weakly held by the positive nucleus.

This has two immediate results. Firstly, the neutral atoms readily lose one (group I) or two (group II) electrons to form ions. Further electrons would have to be removed from the noble gas core, closer to the nucleus, and so would

Element	*Electronic structure*	*1st Ionisation energy* kJmol^{-1}	*Atomic radius/nm*	*Ionic radius/nm (M^+)*	*Heat of atomisation/* kJmol^{-1}	*Standard electrode potential* V
Group I						
Li	(He)$2s^1$	519	0·15	0.06	161	−3.0
Na	(Ne)$3s^1$	494	0.19	0.10	109	−2.7
K	(Ar)$4s^1$	418	0.23	0.13	90	−2.9
Rb	(Kr)$5s^1$	402	0.24	0.15	86	−2.9
Cs	(Xe)$6s^1$	376	0.26	0.17	79	−2.9
Group II						
Be	(He)$2s^2$	2660	0.11	0.03	320	−1.8
Mg	(Ne)$3s^2$	2186	0.16	0.07	150	−2.4
Ca	(Ar)$4s^2$	1740	0.20	0.10	193	−2.8
Sr	(Kr)$5s^2$	1608	0.21	0.11	164	−2.9
Ba	(Xe)$6s^2$	1468	0.22	0.13	176	−2.9

Table 3.1 Properties of group I and II elements.

require much greater amounts of energy; this does not happen. Hence these elements are restricted to a mono- or di-valent state. Secondly, the weak attraction between the outer electron(s) and the nucleus allows the electron(s) to escape to a greater distance, so increasing the size of the atom.(See previous chapter.) This affects any property—such as vaporisation—which depends on separating atoms, since the force between atoms decreases with increased distance.

Within any one group, the size of the atoms increases with atomic number—not unexpectedly—because more electrons have to be accommodated. Several of the more important fundamental properties are given in table 3.1.

3.2 PROPERTIES RELATED TO THE HEAT OF ATOMISATION

The heat of atomisation refers to the process M(s) $\longrightarrow$ M(g) (25 °C) and will be related therefore to other processses involving the separation of atoms. Typical values for other metals are in the range 350–750 $kJmol^{-1}$. By comparison, then, you may expect that group I and II elements have lower boiling points, and also melting points, than usually associated with metals. There is even a sodium–potassium alloy liquid at room temperature.

Deformation also involves separation, and so the above elements are also comparatively soft; in most cases they can be cut with a penknife. This trend can be extended within the groups, lithium being the hardest, and caesium the softest in group I: and all group II metals are harder than those of group I.

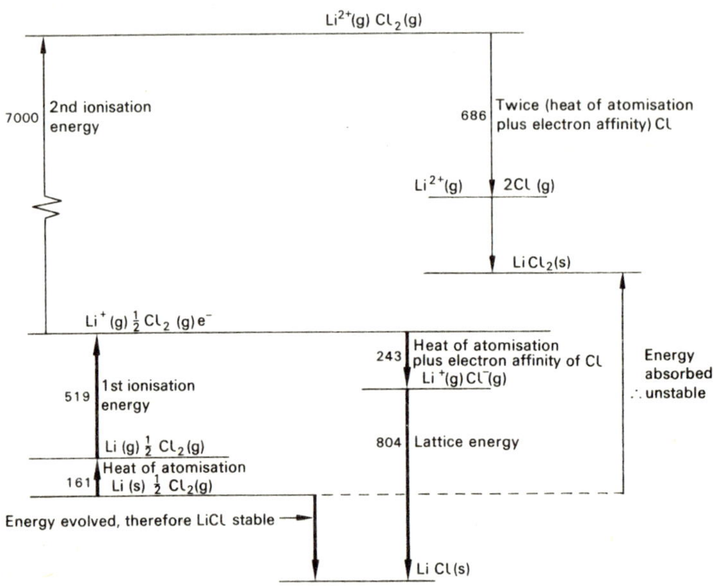

Fig. 3.1 Formation of LiCl *and* $LiCl_2$.

3.3 PROPERTIES RELATED TO IONISATION ENERGY

To convert lithium metal into the lithium ions in LiCl, the lithium atoms can be considered to pass through the stages outlined in the Born–Haber cycle. In this cycle, the endothermic or energy-absorbing stages are $Li(s) \longrightarrow Li(g)$, and $Li(g) \longrightarrow Li^+(g)$, requiring a total of 680 kJmol^{-1} (table 2.1). Unless all this energy and more is repaid, the compound formed will not be stable. In this case, energy is available from the exothermic process, $e^- + \frac{1}{2}Cl_2 \longrightarrow Cl^-$ ($\triangle H = -243\,kJmol^{-1}$) and from the lattice energy, $\triangle H = -840$ kJmol^{-1}. However, if $LiCl_2$ were required, the Li^{2+} ion would have to be formed. Since the second electron is withdrawn from an inner shell, the second ionisation energy is very high, about 7000 kJmol^{-1}. Nothing like this amount can possibly be repaid by either exothermic process above, and therefore lithium is restricted to the chemistry of a monovalent ion. Similar arguments apply to other members of group I and (for divalent ions) of group II.

3.4 PROPERTIES RELATED TO STANDARD ELECTRODE POTENTIAL ($E^\ominus$)

$E^\ominus$ is a measure of the energy of the process $X^+(aq) + e^- \longrightarrow X(s)$. The high *negative* values for elements of group I and II imply that the *reverse* process is likely to occur, so that these metals are good reducing agents (a reducing agent transfers electrons to other substances). The only other element to approach the values given in table 2.1 is aluminium, $E^\ominus = -2.3\,V$ in alkaline solution. Hence all these elements will reduce water to hydrogen.

$$M(s) + H_2O(l) \longrightarrow M^+(aq) + OH^-(aq) + \tfrac{1}{2}H_2(g).$$

In most cases the reactivity within a group follows the values of $E^\ominus$, thus

Sodium	: melts and may ignite	all in
Potassium	: ignites and may explode	cold
Rubidium	: explodes	water

Magnesium	: slow reaction in cold water
Calcium	: fast reaction in hot water
Barium	: fast reaction in cold water.

Lithium is abnormal, possessing a very high value of $E^\ominus$, but reacting more slowly than any other element in group I. This is because the rate of reaction is governed not only by energy output, but also by the activation energy. In order to react, lithium atoms must become detached, or separated. As explained in section 3.2, such stages will be slowest, requiring more energy, in the case of lithium.

Although $E^\ominus$ values refer strictly to aqueous solution, they form a rough guide to behaviour of solids. In particular, the excellence of these metals as reducing agents implies that their ions are difficult to re-oxidise to the metal. Therefore the industrial method of preparation is by electrolysis of a fused salt, usually the chloride.

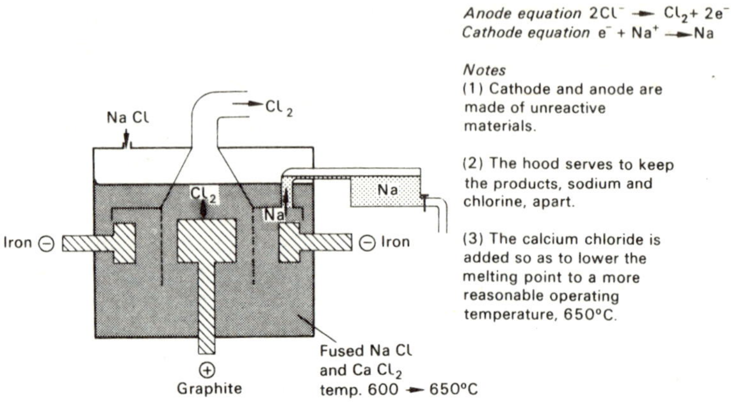

Fig. 3.2 Downs' cell for the manufacture of sodium.

3.5 PROPERTIES DEPENDING ON SIZE OF ION

The number of negative ions that can be packed around a positive ion depends on their relative sizes. The cation is nearly always the smaller of the two, as the removal of an electron allows the remaining excess positive charge to draw in

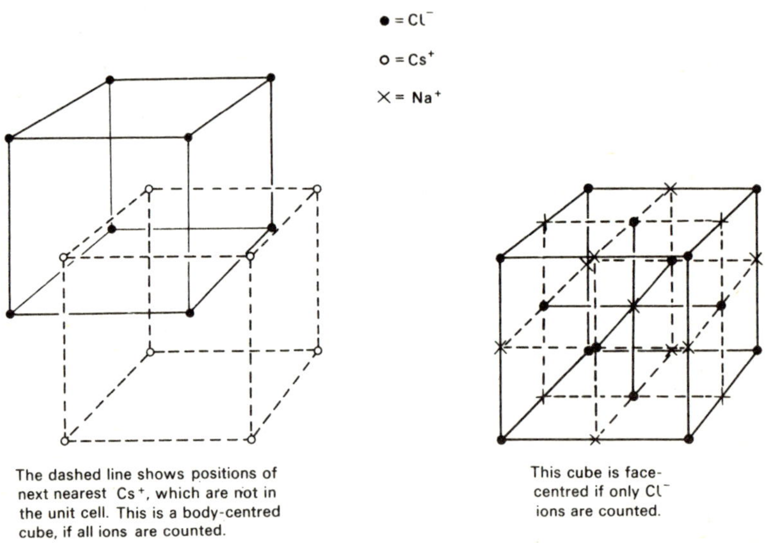

Fig. 3.3 Crystal structures of group I halides.

the remaining electrons even closer. If the cation has at least three-quarters the radius of the anion, eight of the latter can touch the former without touching each other. If this radius ratio (radius M^+/radius Y^-) is between 0.4 and 0.75, it is possible to pack six anions about a cation, and if the ratio decreases to between 0.25 and 0.4 the number decreases to four. (You can experiment in two dimensions by using 2p pieces to surround 2p, 1p and $\frac{1}{2}$p pieces.) For caesium chloride, CsCl, the radius ratio is (0.16/0.18) = 0.9, and this crystallises in a structure of which a unit cell is shown below. Eight chloride ions surround one caesium ion, and since the formula is CsCl, eight caesium ions surround each chloride. This is called 8 : 8 co-ordination. Similarly the radius ratio for sodium chloride is 0.55, and this has a structure with 6 : 6 co-ordination.

3.6 PROPERTIES RELATED TO THE LATTICE ENERGY

The lattice energy is that of the process $M^+(g) + X^-(g) \longrightarrow MX(s)$. As such, it measures the work done when two ions—attracted by an electrostatic force—approach each other until contact is made. Hence lattice energy increases if either the charge on the ion increases, or alternatively, the ionic radius decreases so that the ions can come close to one another. The compounds with the highest lattice energies will therefore contain lithium (if from group I) or beryllium (if from group II), and group II lattice energies are in general much higher than group I.

Consider the process $Li_2O + CO_2 \longrightarrow Li_2CO_3$, compared with $Cs_2O + CO_2 \longrightarrow Cs_2CO_3$. In effect, carbon dioxide is inserted into an existing lattice, causing an expansion. The work required for this expansion is repaid by the energy of combination of oxide ion with carbon dioxide ($O^{2-} + CO_2 \longrightarrow CO_3^{2-}$). However, the work required to expand the lattice is much greater for lithium oxide, with its higher lattice energy, than for sodium oxide or other oxides. Consequently, lithium carbonate will be more unstable than caesium carbonate. The former is, in fact, the only group I carbonate to decompose on heating.

Extension of this argument to group II oxides/carbonates, with yet higher lattice energies, predicts correctly that all are unstable on heating. In the extreme case, beryllium carbonate decomposes even at room temperature.

Similar principles explain the following:

(1) formation of lithium oxide (Li_2O), but sodium peroxide (Na_2O_2) and potassium superoxide (K_2O_4) when the respective elements are burnt in air—insertion of oxygen into an oxide lattice.

(2) instability of group II hydroxides on heating ($Ba(OH)_2 \longrightarrow BaO + H_2O$) compared with stability of group I—insertion of water molecules into an oxide lattice.

(3) stability of polyhalides for potassium (KI_3), rubidium, and caesium ($CsCl_3$), and barium, but not for smaller elements—insertion of halogen, Cl_2, into chloride lattice, etc.

3.7 PROPERTIES RELATED TO LATTICE ENERGY AND HYDRATION ENERGY

When an ionic compound dissolves in water ($M^+X^-(s) \longrightarrow M^+(aq) + X^-(aq)$) the energy change can be represented as the sum of the lattice energy ($M^+X^-(s) \longrightarrow M^+(g) + X^-(g)$) and the hydration energy ($M^+(g) + X^-(g) \longrightarrow M^+(aq) + X^-(aq)$). Unfortunately, both of these quantities increase as ionic radii decrease (see a physical chemistry textbook), the former being endothermic and the latter exothermic, so that an increase in one is cancelled out by an increase in the other.

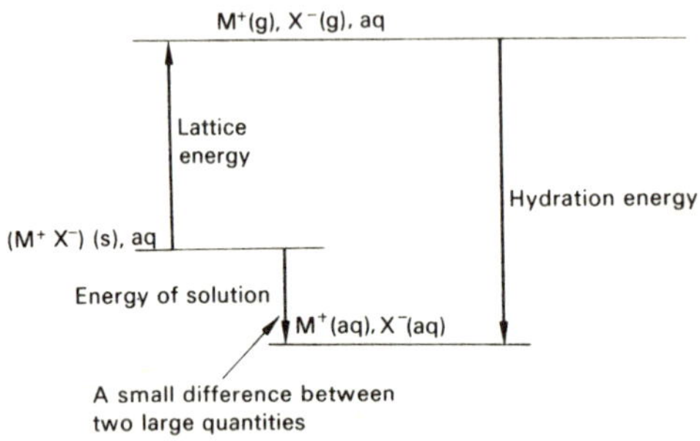

Fig. 3.4 Hydration and lattice energies.

Quantitative predictions about solubility are therefore difficult. The only general rule is that in the set of compounds MX_n, where M is any member of a given group, the least soluble will be the one in which M^+ and X^- are of similar size. Thus, for the large ion sulphate, the compound barium sulphate is insoluble, calcium sulphate is sparingly soluble, but beryllium and magnesium sulphates are soluble. In considering the fluorides of group I, only lithium fluoride, with a similarly small-sized anion and cation, is insoluble. This effect arises because of the higher co-ordination and therefore greater ionic attraction between ions of equal size, as described above.

Water is often combined in crystals as water of crystallisation. Since the negative end of the dipole in water (located near the oxygen atom) can approach most closely to a small ion, then water is most strongly bonded to such ions. The energy is usually enough to offset that required to expand the crystal lattice to accommodate the extra water. (See section 2.6.) Hence the group I salts most often hydrated are those of lithium (e.g. $LiCl.2H_2O$) and sodium (e.g. $Na_2CO_3.10H_2O$, loses water at room temperature). In group II, hydration is more widespread, as higher charge and smaller radius attract the water molecule more strongly (e.g. $MgCl_2$ and $CaCl_2$ are deliquescent, $BaCl_2$ exists as a dihydrate).

3.8 SPECIAL PROPERTIES OF BERYLLIUM

When an ion is formed, the charge is confined to the volume described by the ionic radius. The greater the charge density, the more work will have to be done to create this charge. The value of charge densities are, relative to Be^{2+}, 1; Mg^{2+}, $\frac{1}{8}$; Li^{+}, $\frac{1}{16}$; Na^{+}, $\frac{1}{48}$. Beryllium is obviously exceptional in this respect, and it turns out that Be^{2+} is hardly ever formed.

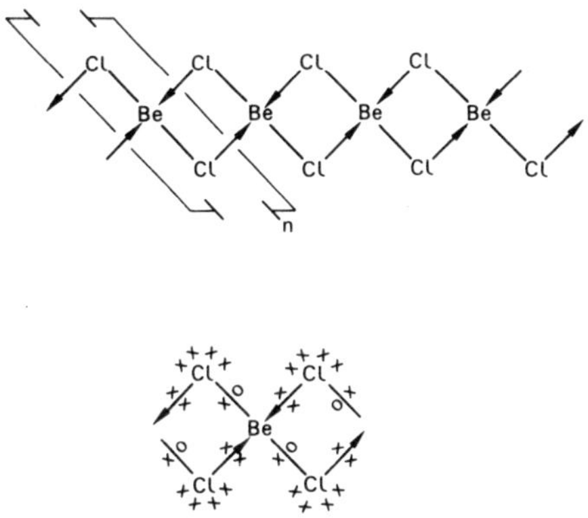

Fig. 3.5 The structure of beryllium chloride.

In anhydrous compounds, beryllium tends to form covalent bonds. For example, $BeCl_2$ sublimes and even when melted under pressure will not conduct electricity. Both these properties are consistent with the polymeric covalent structure found by crystallography (fig. 3.5). Beryllium also forms covalently bonded complexes, such as $Be(OH)_4{}^{2-}$ when it dissolves in alkali (unlike all other s-block elements)

$$Be + 2OH^- + 2H_2O \longrightarrow \underset{\text{tetrahydroxoberyllate(II)}}{Be(OH)_4{}^{2-}} + H_2$$

and is capable of forming a fluoride complex (Other s-block metals form $M^+(aq) + F^-(aq)$).

$$BeO + 4HF \longrightarrow \underset{\text{tetrafluoroberyllate(II)}}{BeF_4{}^{2-}} + H_2O + 2H^+$$

In aqueous solution, it exists as $Be(H_2O)_4{}^{2+}$, the small size preventing the co-ordination of a larger number of water molecules (all other ions of s-block elements co-ordinate six). Even so, the charge is distributed by transferring a proton to a water molecule.

$$Be(H_2O)_4{}^{2+} + H_2O \longrightarrow \underset{\text{tetraaquaberyllium(II)}}{Be(H_2O)_3OH^+} + H_3O^+$$

Beryllium salts are therefore acidic in aqueous solution. Magnesium salts are also acidic but only slightly so. Both magnesium and beryllium chloride solutions emit some hydrogen chloride on evaporation because of this, and varying mixtures of oxide and chloride are left.

3.9 INDUSTRIAL APPLICATIONS OF GROUP I CHEMISTRY

The world production of sodium is about 100 000 tonnes per year. Sodium is used as the element in coolants for nuclear reactions and in lamps; and indirectly as a reagent in making sodium cyanide (for case-hardening of steel) or lead tetra-ethyl (for antiknock petrol).

Sodium hydroxide is made by electrolysis of concentrated brine. The Hooker cell is illustrated in fig. 3.6, and this cell is displacing the older Kastner cell, which requires large amounts of expensive mercury. There are four ions present

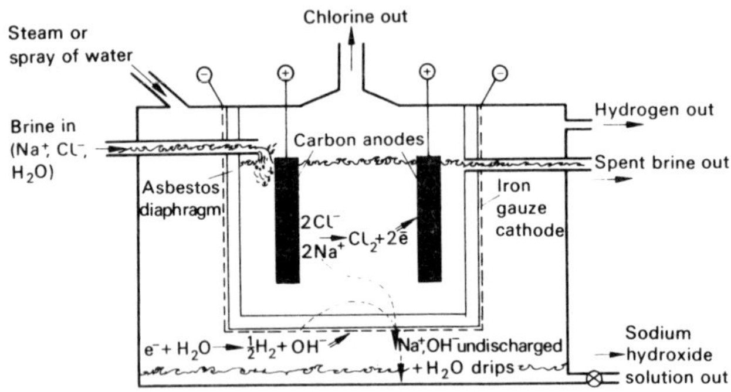

Fig. 3.6 Hooker cell.

in the solution of sodium chloride being electrolysed, namely, Na^+, Cl^-, H^+ and OH^-, some H^+ and OH^- being formed by dissociation of water. Of these, Cl^- and H^+ discharge, to form the valuable by-products chlorine and hydrogen.

$$\textit{Anode} \qquad 2\,Cl^- \longrightarrow Cl_2 + 2e^-$$
$$\textit{Cathode} \qquad H_2O + e^- \longrightarrow \tfrac{1}{2}H_2 + OH^-$$

The discharge of hydrogen ions in the cathode region leaves the solution there alkaline, and the water (condensed steam) dripping from the outside of the gauze cathode bears away a concentrated solution of undischarged sodium and hydroxide ions.

Sodium hydroxide is used as a cheap neutraliser, in the manufacture of rayon, soap and paper, and also in the purification of bauxite. Production in the U.K. alone is about one million tonnes.

Sodium carbonate is the other major product containing sodium. It is produced partly by the reaction of excess sodium hydroxide from the above process with carbon dioxide, and partly by the Solvay, or ammonia–soda, process. Calcium carbonate (limestone) and sodium chloride (rock salt) are ideally cheap as raw materials, but unfortunately are the *products* of the spontaneous reaction

$$Na_2CO_3 + CaCl_2 \longrightarrow CaCO_3 + 2NaCl$$

The complex process represented in the flowsheet is necessary in order to reverse this reaction economically. The calcium compounds never meet the sodium compounds. Instead, sodium chloride, as brine, is saturated first with ammonia, and then with carbon dioxide. The ammonia is a sufficiently strong base to remove the protons from carbonic acid and create a significant concentration of hydrogencarbonate ion (top row, left to right)

$$NH_3 + H_2CO_3 \rightleftharpoons NH_4^+ + HCO_3^-$$

Sodium hydrogencarbonate is only sparingly soluble, and the concentration of hydrogencarbonate is high enough to precipitate it. It is filtered off and heated to produce the carbonate (top right)

$$2NaHCO_3 \longrightarrow Na_2CO_3 + H_2O$$

The calcium carbonate is first heated to release the carbon dioxide needed, and then reacted with the ammonium chloride to regenerate the ammonia (bottom row, right to left)

$$CaCO_3 \longrightarrow CaO + CO_2$$
$$CaO + 2NH_4Cl \longrightarrow CaCO_2 + 2NH_3 + H_2O$$

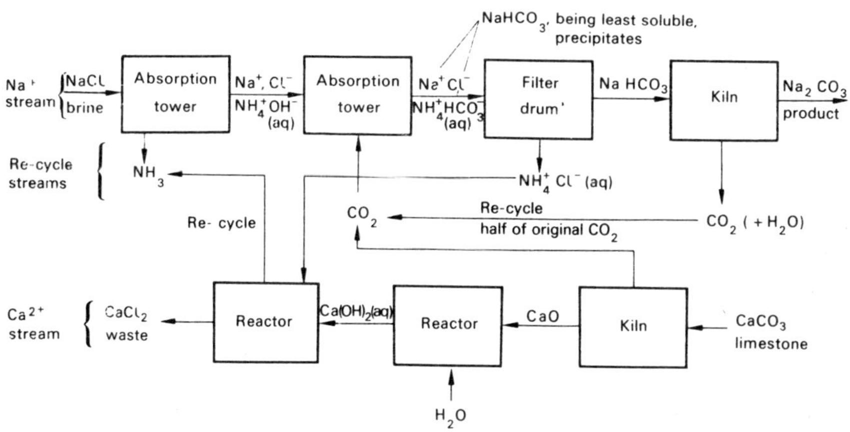

Fig. 3.7 The ammonia–soda process. (i) Diagram of carbonating tower in ammonia–soda process for Na_2CO_3, *showing bubble caps and opposite flow of* $CO_2(g)$ *and ammoniacal brine. (ii) Flowsheet for the ammonia–soda process.*

The recycling of ammonia and carbon dioxide makes the process economical, but difficult to control, and for this reason, the process was patented some years before a successful plant was built. The calcium chloride is a waste product, no significant use being made of it. The sodium carbonate produced is used to soften water by precipitating calcium carbonate. It is also used to make glass, by its reaction with sand.

$$Na_2CO_3 + SiO_2 \longrightarrow Na_2SiO_3 + CO_2$$

Glass is a mixture of silicon(IV) oxide with sodium silicate and other silicates.

Other commercially important group I salts are:

(i) Sodium phosphate (Na_3PO_4), in detergents:
(ii) Sodium nitrite ($NaNO_2$), in dyes and preservatives in food;
(iii) Sodium hydrogencarbonate and hydrogenphosphate ($NaHCO_3$, NaH_2PO_4), in baking powder;
(iv) Sodium sulphite (Na_2SO_3), in the bleaching of paper pulp;
(v) Potassium chloride (KCl) in fertiliser; and
(vi) Potassium nitrate (KNO_3) in gunpowder.

3.10 INDUSTRIAL APPLICATIONS OF GROUP II CHEMISTRY

Magnesium metal is obtained by electrolysing fused magnesium chloride between graphite anodes ($2Cl^- \longrightarrow Cl_2 + 2e^-$) and steel cathodes ($Mg^{2+} + 2e^- \longrightarrow Mg$). The chloride is obtained from the hydroxide and hydrogen chloride. The hydroxide is obtained from sea-water by precipitation of magnesium ions with quicklime (calcium oxide).

$$CaO + H_2O + Mg^{2+} \longrightarrow Mg(OH)_2 + Ca^{2+}$$

The hydroxide can be heated to produce refractory bricks of magnesium oxide

$$Mg(OH)_2 \longrightarrow MgO + H_2O$$

Calcium can be obtained by electrolysis in the same way as magnesium. In this case the calcium chloride is obtained as a by-product of the Solvay process.

Calcium carbonate deposits are widespread throughout the world, and are very valuable if pure. Some calcium carbonate is used in the Solvay process, or mixed with sodium carbonate in glassmaking (see above). Most of the rest is heated to form calcium oxide, which is then slaked with water to form the hydroxide.

$$CaCO_3 \longrightarrow CaO + CO_2 : CaO + H_2O \longrightarrow Ca(OH)_2$$

The hydroxide, known commercially as slaked lime, is often used to correct the pH of the soil, since it is a strong alkali. It is also mixed with sand to form mortar. Mortar sets initially by losing water, and later hardens by re-forming the carbonate.

$$Ca(OH)_2 + CO_2 \qquad CaCO_3 + H_2O$$

Cement is a mixture of calcium hydroxides and calcium aluminium silicates formed by heating limestone and clay. It sets in much the same way as mortar. Calcium sulphate dihydrate ($CaSO_4.2H_2O$) is used as plaster.

3.11 SUMMARY OF COMMON PROPERTIES OF GROUPS I AND II

For additional properties associated with anions, see the summary at the end of the appropriate chapter.

GROUP I	*General property of element*	*Exceptions and gradations*
Physical state	Soft, silvery metal, tarnishing in air	Cs softer than Li and tarnishes more rapidly
Reaction with water	Vigorous or violent reaction $M + H_2O \longrightarrow MOH + \frac{1}{2}H_2$	Lithium slow to caesium violent
Reaction with other elements	Chlorides, bromides, sulphides, hydrides formed on heating	Only lithium forms a nitride
Occurrence	As chlorides	Lithium as aluminosilicate
Halides	All soluble All 6:6 co-ordinated crystal structures (as NaCl)	LiF is insoluble Cs salts are 8:8 co-ordinated
Oxides	Formed by heating in limited oxygen Dissolve to form hydroxides in water $M_2O + H_2O \longrightarrow M(OH)_2$	
Hydroxides	Strong bases, very soluble Alkaline solution in water	LiOH weak base, sparingly soluble.
Carbonates	Soluble, stable to heating, alkaline solution in water	Li_2CO_3 insoluble, decomposes $\longrightarrow$ $Li_2O + CO_2$ on heating Na_2CO_3 forms decahydrate on crystalising, effloresces in air to give monohydrate
Hydrogen-carbonates	Soluble solids Decompose on heating $2NaHCO_3 \longrightarrow Na_2CO_3 + H_2O + CO_2$	$LiHCO_3$, $NaHCO_3$ sparingly soluble
Nitrates	Soluble solids Decompose on heating $MNO_3 \longrightarrow MNO_2 + \frac{1}{2}O_2$	Lithium nitrate forms oxide on heating $2LiNO_3 \longrightarrow Li_2O + 2NO_2 + \frac{1}{2}O_2$ (see chapter 6)
Sulphates	Soluble solids	Sodium, lithium salts hydrated

GROUP II	*General property of element*	*Exceptions and gradations*
Physical state	Hard grey metals, tarnish in air	Barium softest, and tarnishes rapidly Magnesium, beryllium only form surface coating
Reaction with oxygen	Burn at high temperatures to form oxide $M + \frac{1}{2}O_2 \longrightarrow MO$	Barium forms peroxide BaO_2
Reaction with water	Form hydroxide $M + 2H_2O \longrightarrow M(OH)_2 + H_2$	Beryllium hardly reacts, magnesium with hot water or steam, the rest increasingly rapid with cold water
Reaction with other elements	Form halides, hydrides, sulphides, nitrides on heating	MgH_2, BeH_2 do not exist
Occurrence	Often as carbonates, sometimes sulphates	Beryllium as complex oxide with Al, Si: barium mainly as sulphate
Halides	Fluorides insoluble others soluble and hydrated Evaporation of chloride solution to dryness yields oxide $MCl_2 + H_2O \longrightarrow MO + 2HCl$	BaF_2 most soluble. $BeCl_2$ covalent if anhydrous $CaCl_2$, $MgCl_2$ deliquescent, hexahydrates: $BaCl_2$, dihydrate. $BaCl_2$ can be dehydrated by heating
Hydrides	Ionic solids $M^{2+}(H^-)_2$	Not formed by Mg, Be
Oxides	React with water to form alkaline solutions $MO + H_2O \longrightarrow M(OH)_2$	BeO insoluble, MgO hardly reacts, others vigorous
Hydroxides	Strong bases, sparingly soluble	$Be(OH)_2$ weak base and insoluble, solubility increases towards Ba
Carbonates	Insoluble, decompose on heating $MCO_3 \longrightarrow MO + CO_2$	Be most unstable (decomposes 20 °C) Ba most stable (decomposes 1200 °C)
Hydrogen-carbonates	Stable only in solution, decompose on evaporation to form carbonate $M(HCO_3)_2 \longrightarrow MCO_3 + H_2O + CO_2$	For reactions in carbon cycle and hard water see Chapter 5 and see also below
Nitrates	Deliquescent, unstable to heat $M(NO_3)_2 \longrightarrow MO + 2NO_2 + \frac{1}{2}O_2$	$Ba(NO_3)_2$ not deliquescent
Sulphates		Solubility decreases from Be, Mg (soluble) to Ba (insoluble)
Hard water	Due to Ca^{2+}, Mg^{2+} ions in water from (a) solution of sulphates (b) chemical reaction of carbonates with carbon dioxide in ground water $MCO_3 + H_2CO_3 \longrightarrow M(HCO_3)_2$ Disadvantages of hard water	'Permanent', can only be removed by adding Na_2CO_3($Ca^{2+} + CO_3^{2-} \longrightarrow CaCO_3$) or ion-exchange resin 'Temporary', reaction can be reversed on heating when carbonate deposits form in boilers Scum formed with soap sodium stearate converted to calcium stearate, $Ca^{2+} + 2\,(\text{stearate}^-) \longrightarrow Ca(\text{stearate})_2$) Boilers need de-scaling

QUESTIONS

1 Show how the elements Li–Cs possess similarities and yet also form a graded series. Actual observable properties should be quoted as well as general trends. Alternatively, do the same for the series Be–Ba.

2 From data and trends given in the chapter, predict the following properties of $_{87}Fr$ and $_{88}Ra$: electronic structure, ionic radius, heat of atomisation, hardness, rate of reaction with water, solubility of sulphate, thermal stability of carbonate, and occurrence of hydrated salts. Also predict any other three properties of your choice.

3 Describe the process by which sodium can be obtained from common salt.
(a) Are there any other materials which could be used for electrodes in this process? What restrictions are there on the choice of such materials?
(b) Suggest, with reasons, what additives might be used in place of calcium chloride.
(c) If you, as manufacturer, were primarily interested in chlorine production, would you choose to electrolyse molten sodium chloride or aqueous sodium chloride? Justify your choice.
(d) Devise, and draw a diagram of, a cell suitable for the production of strontium metal.

4 (a) The flow sheet for the ammonia–soda process is shown in fig. 3.7 in such a way as to emphasise the conversions $NaCl \longrightarrow Na_2CO_3$ (top line) and $CaCO_3 \longrightarrow CaCl_2$ (bottom line). Redraw it to show the recycling of carbon dioxide and ammonia (e.g. draw two circles and link in the other processes), thus

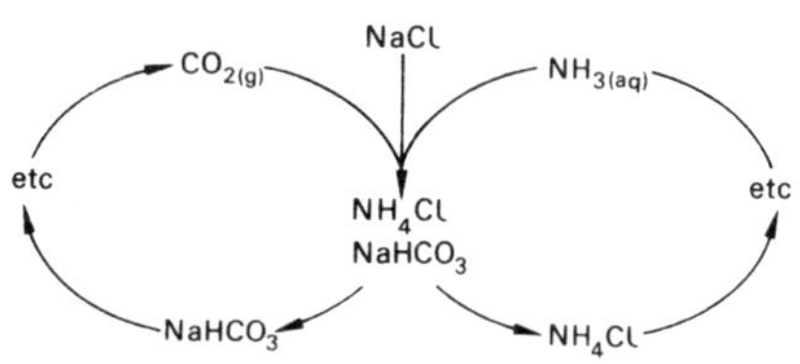

(b) Label the above diagram with balanced equations.
(c) Some sodium carbonate is now made by blowing carbon dioxide into sodium hydroxide. How is the emergence of this process connected to the present heavy demand for chlorine? (How is chlorine made?)

5 Explain the following observations (some explanations are included in the chapter, some are not).
(a) On burning the metals in air, lithium forms an oxide, sodium a peroxide and potassium a superoxide.

(b) The radioactive $_{90}Sr$ is particularly dangerous among radio-isotopes, causing bone marrow damage.
(c) Lithium forms more hydrated simple salts, than the rest of group I put together.
(d) Beryllium chloride does not conduct electricity when molten.
(e) Calcium chloride, although very cheap commercially, is not used to melt ice on roads in winter (this is connected with rust damage).
(f) Although lithium and beryllium both have high hydration energies, their redox potentials differ, being in one case the highest in the group, and in the other the lowest.
(g) Ionic compounds of the formula M^+X^- need not have the same crystal structures.
(h) Barium chloride is intensely poisonous, but barium sulphate is swallowed routinely before X-ray examination of the stomach.
(i) The hardness and boiling point of a metal are loosely connected.
(j) Calcium hydroxide is sold to farmers.

6 The following statements are not correct in all respects. There may be exceptions, ambiguities, omissions or deliberate mistakes. Write out a correct version of the statement, and comment on the theoretical background to it.
(a) Calcium(I) fluoride is unstable.
(b) The atomic radius of elements in group II is directly proportional to the atomic number.
(c) Sodium chloride conducts electricity.
(d) All group I salts are colourless.
(e) The general formula for a group II ion in water is $M(H_2O)_6^{2+}$.
(f) Caesium exists as Cs^{3+} in $CsCl_3$.
(g) The fourth ionisation energy of lithium is 10000 $kJmol^{-1}$.
(h) Sodium hydrogencarbonate forms an acidic aqueous solution.
(i) The cathode process in the electrolysis of brine is $Na^+ + e^- \longrightarrow Na$.
(j) The large ionisation energy of lithium means that its salts are less stable.

4

Aluminium and related elements

A group in which relations are governed by charge/radius ratio

4.1 INTRODUCTION

Aluminium is the most abundant metal in the earth's crust. Although the principal ore is bauxite, $Al_2O_3.2H_2O$, tremendous amounts of aluminium occur as hydrated potassium aluminium silicates forming clays, slates, shales and other rocks. At the present time it is uneconomic to obtain aluminium from these sources.

The chemistry of aluminium is dominated by the high charge (3+) and small radius (0.05 nm) of its ion. The combined effect of these two factors is expressed as the ratio charge/radius. Some values are given in table 4.1. The value obtained is important because it measures the energy released when one of these ions approaches and comes into contact with an ion of opposite charge. (Strictly, if two ions of charge z_1, z_2, approach to within a distance r, having come from infinity, work exchanged equals $z_1z_2/4\pi\varepsilon^0r$.)

$Li^+ \frac{1}{0.06} = 16.6$	$Be^{2+} \frac{2}{0.03} = 66$	$B^{3+} \frac{3}{0.02} = 150$	–
$Na^+ \frac{1}{0.095} = 10.5$	$Mg^{2+} \frac{2}{0.066} = 30$	$Al^{3+} \frac{3}{0.05} = 60$	$Si^{4+} \frac{4}{0.04} = 100$

Table 4.1 Values of charge/radius ratio (elementary charge nm^{-1}).

4.2 COMPETITION BETWEEN COVALENT BONDS AND IONIC BONDS

If aluminium ions, with their high charge and low radius, are in contact with anions, the electrons of the anion will be distorted (polarised) by the strong electric field of the aluminium. As soon as there is an appreciable electron density between the aluminium and the anion, we can say that there is a degree of covalent bonding (II, fig. 4.1). In extreme cases, further polarisation may allow a fully covalent bond to develop (III, fig. 4.1).

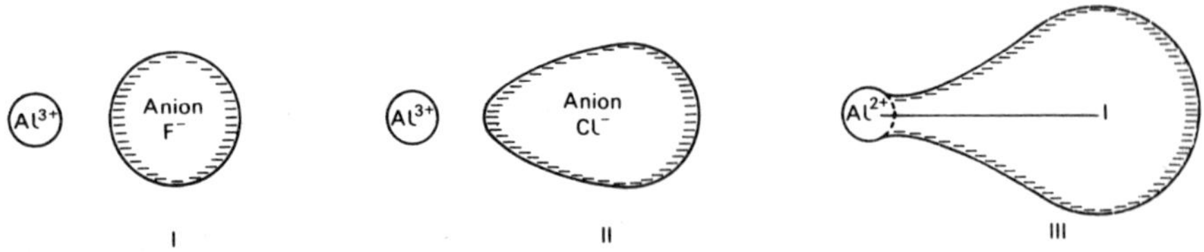

Fig. 4.1 Polarisation of anions.

These extreme cases will include large anions, since the nucleus of such an ion has little attraction for the electrons in its outer shell.

A series of examples is provided by the aluminium halides. Aluminium fluoride is ionic and has the symmetrical crystal structure shown in fig. 4.2, since the fluoride ion is small and non-polarisable. Aluminium bromide and iodide, with their large anions, are covalent. In fact, these have a dimeric structure, in order that each aluminium atom can achieve an outer octet of electrons, by using one non-bonding pair of electrons from a bromine atom to form a co-ordinate bond.

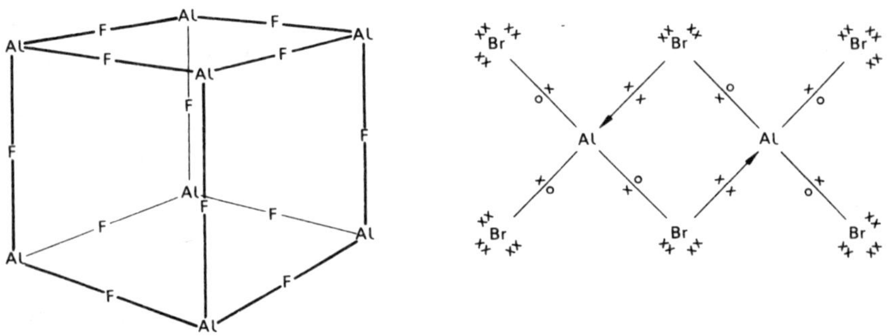

Fig. 4.2 Structures of aluminium halides.

Aluminium chloride occupies an intermediate position. The solid exists as a partly symmetrical crystal, indicating bonds intermediate between ionic and covalent. It sublimes to yield the covalent molecules Al_2Cl_6 with the same structure as Al_2Br_6. At higher temperatures these dissociate to $AlCl_3$. Similarly Al_2Br_6 dissociates to $AlBr_3$ when heated. The degree of ionic bonding is reflected in the melting points: AlF_3, 1260 °C; $AlCl_3$, 180 °C (sublimes); Al_2Br_6, 98 °C; Al_2I_6, 180 °C.

4.3 FORMATION OF STRONG BONDS TO OXYGEN

Oxide ions carry two negative charges, as opposed to the halide's one. Thus, the oxide ion is yet more polarisable and only forms covalent bonds with aluminium. Since both Al^{3+} and O^{2-} ions are small, the bond is very strong.

Paradoxically, this is responsible for both the reactivity and unreactivity of aluminium!

Aluminium is familiar from its use in household articles such as saucepans, in lightweight car and lorry bodies, and in aircraft fuselages. All these uses are possible because the metal is coated with a film of oxide, covalently bound, and about 1 nm thick. The oxide is covalent, not ionic, and so it is resistant to acid attack:

$$2H^+ + O^{2-} \longrightarrow H_2O$$

Since it is a giant molecule, with strong bonds, it is resistant to abrasion, and in any case it re-forms quickly if damaged. Hence aluminium is riveted, or has to be welded in an inert atmosphere of argon. Often the oxide film is thickened by using the aluminium article as the anode, at which oxygen is released, in the electrolysis of sulphuric acid (anodising). The protection given by this strongly adherent film is in strong contrast to the process of corrosion of iron, with its flaky coating of rust.

Once the film is weakened, aluminium will react rapidly with many reagents. It will burn in air or steam above about 600 °C to give the oxide.

$$2Al + 3H_2O \longrightarrow Al_2O_3 + 3H_2$$

It will dissolve in acid or alkali to form oxo-complexes. The reaction needs warming to begin with, but often gets out of hand thereafter.

$$Al + 3H^+ + 6H_2O \xrightarrow[\text{HCl}]{\text{hot, conc.}} \underset{\text{hexaaquaaluminium(III)}}{Al(H_2O)_6{}^{3+}} + \tfrac{3}{2}H_2$$

$$Al + 3OH^- + 3H_2O \xrightarrow[\text{NaOH}]{\text{hot, dil.}} \underset{\text{hexahydroxoaluminate(III)}}{Al(OH)_6{}^{3-}} + \tfrac{3}{2}H_2$$

The surface oxide film may also be weakened by washing a piece of aluminium foil in mercury(II) chloride. Mercury is displaced from this salt by aluminium and forms an amalgam (alloy) with aluminium underneath. The oxide does not adhere to the amalgam, and, if the foil is left in air, 'feathers' of aluminium oxide grow from its surface.

4.4 POLARISATION OF NEIGHBOURING O–H BONDS

Aluminium salts in water are strongly hydrated, forming the hexaaquaaluminium(III) ion, $Al(H_2O)_6{}^{3+}$. Owing to the polarising ability of the highly charged Al^{3+} in the centre of this complex, the electrons in the O–H bond are withdrawn in that direction. Consequently hydrogen ions are easily captured from the complex as a proton by any base present. In aqueous solution, therefore, the water acts as base, hydroxonium ions are formed and the solution becomes quite strongly acidic.

$$Al(H_2O)_6{}^{3+} + H_2O \rightleftharpoons Al(H_2O)_5OH^{2+} + H_3O^+$$

This reaction is the first step in a continuous sequence, in which six protons may be lost:

	$Al(H_2O)_6^{3+}$ hexaaquaaluminium (III)	Very weak base e.g. H_2O	Weak base e.g. CO_3^{2-}, S^{2-}	Strong base e.g. OH^-
↑ $+H^+$	⇅ $Al(H_2O)_5(OH)^{2+}$ ⇅ $Al(H_2O)_4(OH)_2^+$ ⇅ $Al(H_2O)_3(OH)_3$ hydrated aluminium (III) hydroxide	↓ i.e. aluminium salts are acidic in solution	↓ i.e. aluminium *hydroxide* is precipitated by carbonates (to yield CO_2).	↓ hydroxide is formed as white precipitate
$-H^+$ ↓	⇅ $Al(H_2O)_2(OH)_4^-$ ⇅ $Al(H_2O)(OH)_5^{2-}$ ⇅ $Al(OH)_6^{3-}$ hexahydroxo-aluminate(III)		Note that $Al(H_2O)_3(OH)_3$ is insoluble because it is the only uncharged species in the sequence	↓ but then re-dissolves in excess hydroxide as more protons are removed and an anion is formed.

The equations for the above reactions should be written

$$3HCO_3^- + Al(H_2O)_6^{3+} \rightleftharpoons Al(H_2O)_3(OH)_3 + 3H_2O + 3CO_2$$

or

$$6OH^- + Al(H_2O)_6^{3+} \rightleftharpoons Al(OH)_6^{3-} + 6H_2O$$

thereby showing the proton transfer from the hydrated aluminium(III) ion to the base.

Since the hydrated aluminium(III) hydroxide can both accept and lose protons, it will react with acids as well as bases. This type of behaviour is said to be amphoteric, and is put to use in the extraction of aluminium, as the bauxite ore is purified.

Crude bauxite, $Al_2O_3.2H_2O$, contains as impurities, iron(III) oxide, titanium(IV) oxide, and silicon(IV) oxide. On treatment with concentrated sodium hydroxide, the aluminium and silicon oxides dissolve (as $Al(OH)_6{}^{3-}$ and $SiO_3{}^{2-}$) leaving the iron(III) and titanium(IV) oxides behind, since the Fe^{3+} and Ti^{4+} ions are larger and do not form such strongly bonded complexes.

After filtering, dilution with water precipitates aluminium(III) hydroxide, which is heated to form aluminium(III) oxide, and used in electrolysis as below. The silicate is not precipitated owing to the greater acidity of the silicate ion which will not accept protons from water.

4.5 EXTRACTION

Aluminium has a very high redox potential, caused by the energy of formation of the Al–O bonds in $Al(H_2O)_6{}^{3+}$. Only the oxide film previously referred to prevents it from reducing water.

$$Al^{3+}(aq) + 3e^- \longrightarrow Al \qquad E^\ominus = -1.66\,V$$
$$H^+(aq) + e^- \longrightarrow \tfrac{1}{2}H_2 \qquad E^\ominus = 0.0\,V$$

An attempted reduction of aqueous aluminium ions would be more likely to reduce the water to hydrogen, therefore. Extraction under anhydrous conditions is more promising.

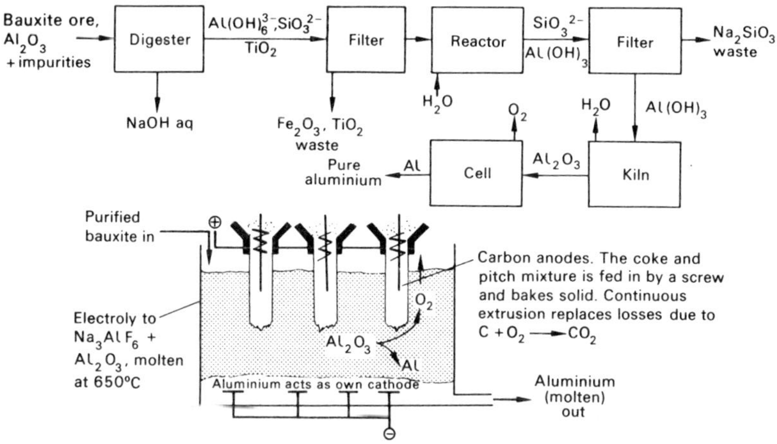

Fig. 4.3 Aluminium extractions.

The oxide, which is readily available, has such a high heat of formation (strong Al–O bonds again) that it can be chemically reduced only with such reagents as sodium metal. This was done in the nineteenth century, but the price of aluminium was nearly the same as that of gold! Electrolysis also presents problems, since the only truly ionic compound of aluminium is the fluoride. Fortunately, the oxide is very soluble in the molten fluoride (cryolite, Na_3AlF_6, is used for convenience), and it is possible to carry out the electrolysis at about 650 °C.

Although the sodium and fluoride ions carry the current, the electric field at the electrodes is sufficient to dissociate the covalent aluminium oxide. The aluminium and oxide ions formed are discharged at the cathode and anode respectively, leaving the opposite ion in the melt.

Anode $Al_2O_3 \longrightarrow 2Al^{3+} + \frac{3}{2}O_2 + 6e^-$
Cathode $6e^- + Al_2O_3 \longrightarrow 2Al + 3O^{2-}$

This case is similar to the electrolysis of sodium sulphate in water, where sodium and sulphate ions carry the current, but the covalent molecule water dissociates and discharges.

Anode $H_2O \longrightarrow 2H^+ + \frac{1}{2}O_2 + 2e^-$
Cathode $e^- + H_2O \longrightarrow \frac{1}{2}H_2 + OH^-$

4.6 RESEMBLANCE TO NEIGHBOURING ELEMENTS

Aluminium resembles a number of elements around it in the periodic table for a number of different reasons. The similarities fall into three groups:

(1) similarity in electronic structure and in size,
(2) similarity in electronic structure but not in size,
(3) similarity in charge/radius ratio.

4.6.1. **Similarity in electronic structure and size**

The group showing similarity in electronic structure and size to aluminium includes gallium, indium and thallium, all elements in group III of the periodic table.

Element	*Electronic structure*	*Atomic radius*/nm	*Ionic radius*/nm
Al	2:8:3	0.14	0.05
Ga	2:8:18:3	0.14	0.06
In	2:8:18:18:3	0.16	0.08
Tl	2:8:18:32:18:3	0.17	0.09

These elements form a range of compounds similar to aluminium, such as dimerised chlorides, X_2Cl_6; hexaaqua complex ions, $X(H_2O)_6{}^{3+}$, which are acidic in solution; thus the metals dissolve in acid and (except Tl) alkali.

The differences are mainly of degree. As the ionic radius becomes larger, features which depend on the charge/radius ratio become weaker. Thus, solutions of thallium(III) ions are not so acidic as those of aluminium(III) ions; thallium(III) oxide is ionic and acts as a base, whereas aluminium oxide is amphoteric; the redox potentials X^{3+}/X decrease, and the metals can be obtained by electrolysis of aqueous solutions.

In addition, thallium shows the 'inert pair' effect, discussed more fully in the next chapter. Thallium(I) salts are common, and indeed formed by the decomposition of thallium(III) halides on heating. Gallium and indium also form compounds in oxidation state I, often the iodide or sulphide. Aluminium appears to have only aluminium(I) chloride as its representative, present in equilibrium at high temperatures

$$2\,Al + AlCl_3 \overset{600\ ^\circ C}{\rightleftharpoons} 3\,AlCl$$

4.6.2 **Similarity in electronic structure but not in size**

This category refers to boron. It is not possible to obtain enough energy from a chemical reaction to concentrate three charges in a sphere of 0.02 nm radius. Boron, therefore, differs from aluminium in that it never forms an ion such as B^{3+}. The polarising effect (section 4.2) is much greater, so that the (B–O)–H bond dissociates readily. In addition, the valence shell involved in bonding can take a maximum of only eight electrons in boron (aluminium, eighteen). Hence six-co-ordinate complexes cannot be formed, since each co-ordinate bond would provide two electrons.

The following series illustrates these and other points.

BCl_3	Liquid, never dimerised, probably too difficult to pack 4 Cl atoms round boron	Ionic solid, dimerises readily to covalent molecule in vapour	$AlCl_3$
H_2O ↓			H_2O ↓
$B(OH)_3 + 3\,HCl$	(1) 3 bonds to boron (2) Strongly acidic solution due to complete hydrolysis and formation of HCl	(1) 6 bonds to aluminium (2) Weakly acidic solution due to partial dissociation of cation	$Al(H_2O)_6{}^{3+} + 3\,Cl^-$
heat ↓			heat ↓
B_2O_3	Covalent oxide, soluble in alkalis only	Covalent oxide, soluble in either acid or base	Al_2O_3

4.6.3 **Similarity due to charge/radius ratio**

At the beginning of this chapter, it was shown how the charge/radius ratio governed the tendency to form covalent bonds, the acidity of hydrated cations, etc. Elements possessing a similar value for the charge/radius ratio should therefore have some properties in common. In particular table 4.1 predicts a considerable similarity between aluminium and beryllium. In practice, the following features are noted.

Beryllium	*Resemblance to aluminium*	*Difference from group II*
Metal is soluble in alkali $Be + 2OH^- + 2H_2O \rightarrow Be(OH)_4^{2-} + H_2$	Metal also dissolves in alkali $Al + 3OH^- + 3H_2O \rightarrow Al(OH)_6^{3-} + \frac{3}{2}H_2$	Metals do not dissolve in alkali
$BeCl_2$ is covalent with 4 co-ordinate structure 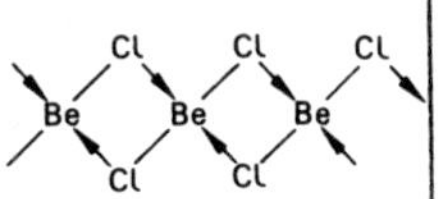	$AlCl_3$ is covalent in vapour with 4 co-ordinate structure 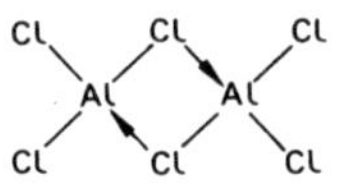	Halides have high co-ordination typical of ionic solids
Be^{2+} does not exist	Al^{3+} probably exists in fluoride only	Many ionic solids
$Be(H_2O)_4^{2+}$ is acidic in solution	$Al(H_2O)_6^{3+}$ is acidic in solution	Slight acidity observed for $Mg(H_2O)_6^{2+}$ only
$BeCO_3$ is unstable at room temperature	$Al_2(CO_3)_3$ is unknown	Carbonates stable up to 800°C or more
$Be(OH)_2$ is amphoteric	$Al(OH)_3$ is amphoteric	All hydroxides are basic

This resemblance is called a 'diagonal relationship'. The comparison between beryllium and aluminium is most striking, but there are similar, though less obvious, likenesses between lithium and magnesium. The relative values of charge/radius ratio in table 4.1 also suggest that the points of comparison should be fewer.

Thus, lithium resembles magnesium but differs from group I in that:

It forms some insoluble salts (Li_2CO_3, LiF).
It reacts directly with nitrogen ($\longrightarrow Li_3N$).
Some salts are thermally unstable (LiOH, Li_2CO_3).

Looking in the opposite direction in the period, there should also be resemblances between boron and silicon. However, the principle of the charge/radius ratio cannot be expected to hold when neither element forms charged ions. There are, however, a few similarities between boron and silicon.

Boron and silicon halides are completely hydrolysed by water.
Boron and silicon hydrides are gaseous and spontaneously inflammable.
Boron and silicon oxides are sufficiently acidic to dissolve metal oxides forming borate or silicate glasses.

4.7 SUMMARY OF REACTIONS OF ALUMINIUM

The reactions of aluminium and its compounds are summarised in fig. 3.4.

Alums, the well-known octahedral crystals, are omitted from the chart. They have the formula $M^IM^{III}(SO_4)_2 12H_2O$, where M^I may be K^+, Rb^+, NH_4^+ and M^{III} may be Al^{3+}, Cr^{3+}, Fe^{3+}. They can be prepared by crystallising equimolar quantities of any two sulphates (e.g. $K_2SO_4 + Al_2(SO_4)_3$). The crystal structures are so similar that mixed crystals containing all the ions may be grown, or a crystal of one alum (e.g. $KAl(SO_4)_2$) will serve as seed crystal for another.

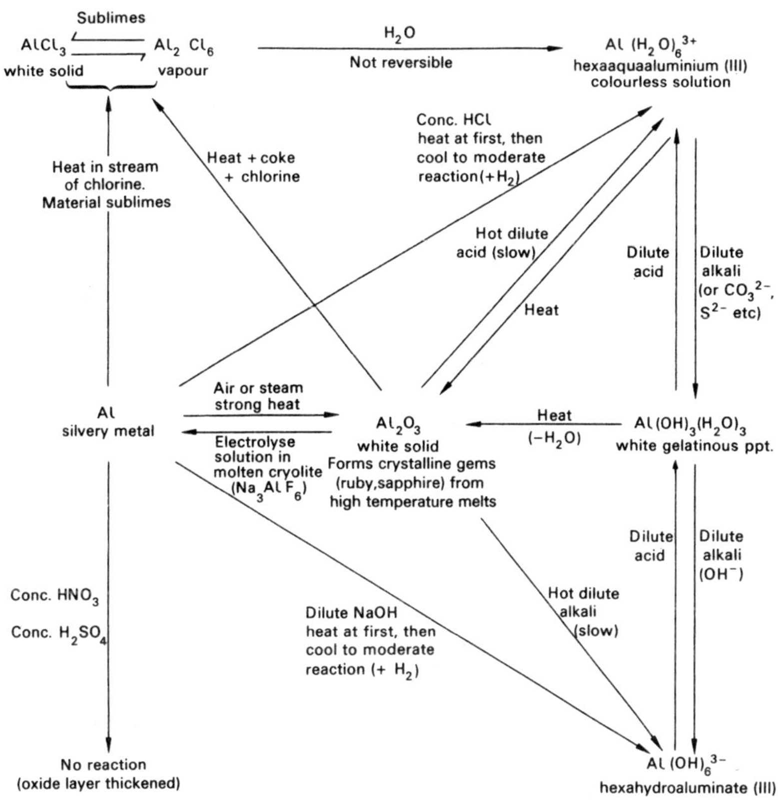

Fig. 4.4 Reactions of aluminium.

Substance	*Uses*
Aluminium metal	Many low density alloys
Aluminium oxide	Catalyst in oil cracking; refractory bricks, cement
Aluminium chloride	Catalyst in benzene reactions
Aluminum sulphate	Coating paper, coagulant (precipitating organic material from water, clotting blood)
Aluminium hydroxide	Mordant (binds dyes to cloth)

Table 4.2 Uses of aluminium and its compounds.

The uses are related to some of the properties discussed. Strong Al–O bonds allow the use of alumina (aluminium oxide) in refractory bricks, and also enable aluminium hydroxide to co-ordinate to cellulose fabrics and to oxygen atoms in dyes.

The coagulating properties of Al^{3+} arise because the fine suspensions of humus (in water) and cells (in blood) are maintained by mutual repulsion of negative charge on the particles. Aluminium ions attract and form clumps with these, which are heavy enough to settle out.

QUESTIONS

1 How is aluminium related in its atomic and electronic properties to (a) beryllium (b) boron (c) gallium? Cite some observable properties and trace the dependence of these to the theoretical relationship first quoted.

2 Describe, briefly, the extraction of aluminium from bauxite. Comment on the following (the answer, if not contained in the chapter, is of a common-sense nature).
(a) Foreign exchange is expended in buying bauxite (or processed metal) but widespread deposits of clay, containing aluminium, in this country are untouched.
(b) Aluminium smelters are frequently sited in rainy, mountainous regions.
(c) Aluminium is rarely used where articles have to be welded.
(d) Iron and copper, though rarer, were used by man long before aluminium.
(e) The oxygen liberated along with the aluminium is not considered suitable for sale for medical or other purposes.

3 (a) What is meant by a diagonal relationship?
(b) Why should such relationships exist?
(c) Why are such relationships restricted to (i) first and second row elements, (ii) the left-hand side of the periodic table?

4 Explain the following observations:

(a) Aluminium has a very high redox potential, yet can safely be heated on a red-hot stove.

(b) The precipitate formed on the addition of sodium sulphide solution to aluminium chloride solution contains no sulphur.

(c) Molten aluminium bromide does not conduct electricity but an aqueous solution of it will do so.

(d) $Al(H_2O)_3(OH)_3$ is less soluble than either $Al(H_2O)_4(OH)_2^+$ or $Al(H_2O)_2(OH)_4^-$.

(e) The molecular mass of aluminium chloride vapour, as determined by vapour density measurements, is 267 at 180 °C and 133.5 at 600 °C, but the latter figure decreases sharply in the presence of solid aluminium.

(f) BF_3 is a gas (b.pt. −100 °C) but AlF_3 is a solid (m.pt. 1260 °C).

(g) On being dissolved in excess alkali, beryllium forms $Be(OH)_4^{2-}$ but aluminium forms $Al(OH)_6^{3-}$ (cf. BCl_3/Al_2Cl_6).

(h) Aluminium sulphate solution gives off hydrogen in the presence of magnesium.

(i) Thallium(III) oxide does not dissolve in alkalis but aluminium(III) oxide does.

(j) Aluminium burns more readily in chlorine than in oxygen.

5 Identify the following elements and compounds, account for the data given, and write equations for the changes occuring.

(a) 1g of element A dissolves in dilute hydrochloric acid giving off 2.5 litres of hydrogen. The solution, when sodium hydroxide solution was added, gave a white precipitate which dissolved in excess alkali.

(b) 1 g of element B reacts with 0.93 g of chlorine. The chloride formed conducts electricity when molten, and 38.3 g of the element are released by one Faraday of electricity.

(c) 0.391 g of a liquid C reacted vigorously with water to form an acidic solution which required 100 cm³ of $\frac{M}{10}$ sodium hydroxide for neutralisation. The original compound also reacted to form an adduct with ammonia in a 1 : 1 mole ratio.

6 The following statements are not correct in all respects. There may be exceptions, ambiguities, omissions or deliberate mistakes. Write out a correct version of the statement and comment on the theoretical background to it.

(a) Aluminium does not form ionic compounds.

(b) Group III halides occur as the dimers X_2Cl_6.

(c) Diagonal relationships apply to first and second row elements only.

(d) Beryllium chloride has a molecular mass of 80.

(e) Aluminium nitrate solution contains H^+ ions.

7 Write balanced equations for

(a) aluminium $\xrightarrow{\text{dilute acid}}$ hexaaquaaluminium(II) chloride.

(b) aluminium $\xrightarrow{\text{dilute alkali}}$ sodium hexahydroxoaluminate(III).

(c) hexaaquaaluminium(III) $\xrightarrow{\text{carbonate}}$ aluminium hydroxide.

5

Group IV: Carbon, Silicon, Germanium, Tin, Lead

A group in which trends overshadow similarities

5.1 INTRODUCTION

This group of elements, situated in the middle of the periodic table, provides excellent examples of trends continuing and developing from one end of a group to the other. Although every element bears a considerable similarity to its neighbours—enough for the properties of germanium to be accurately forecast in advance of its discovery—the first and last elements, carbon and lead, have little in common.

5.2 PROPERTIES DEPENDING ON ELECTRONIC STRUCTURE

All these elements possess an outer shell of four electrons, of which two are in a slightly higher energy level (the p sub-shell) than the other two (in the s sub-shell). The removal of an electron from a positive ion requires progressively more energy as the positive charge increases. Thus the successive ionisation energies of carbon are 1100, 2350, 4600 and 6200 $kJ mol^{-1}$ respectively. This substantial increase is too large to be recovered by lattice energy, so that the tetravalent compounds are typically covalent, and exist as individual molecules. Consequently, in this group, all the tetrahalides are volatile liquids, and all the tetrahydrides are gases.

In some divalent tin and lead compounds, ions do exist, and in these cases, the p electrons have been lost.

$$Sn[\equiv Kr + 4d^{10}5s^2 5p^2] \longrightarrow Sn^{2+}[\equiv Kr + 4d^{10}5s^2] \quad 5$$
$$Pb[\equiv Xe + 4f^{14}5d^{10}6s^2 6p^2] \longrightarrow Pb^{2+}[\equiv Xe + 4f^{14}5d^{10}6s^2]$$

Superficially, the p electrons have been lost because of their lower ionisation energy compared to the s electrons, but this is only partly true, since other terms in the Born–Haber cycle, such as lattice energy, must be considered. Not only must the energy of formation of, say, $SnCl_2$ from its elements be favourable, but also the reaction

$$\underset{\text{(covalent)}}{SnCl_2} \longrightarrow \underset{\text{(ionic crystal)}}{Sn^{2+}\ 2Cl^-}$$

must go forward, otherwise the substance will be a covalent liquid. Cycles for both processes are shown in fig. 5.1. It can be seen, on examining the figures, that the critical factor is the lower ionisation energies of tin and lead. The higher values possessed by carbon, silicon and germanium make the formation of X^{2+} improbable, and the bonding of XCl_2, if it does exist, is covalent. The stability of the state in which the p electrons are involved to the exclusion of the s electrons is called the *inert pair effect*. It occurs in other groups besides group IV: for example thallium (group III) forms Tl^+ and bismuth (group V) forms Bi^{3+}.

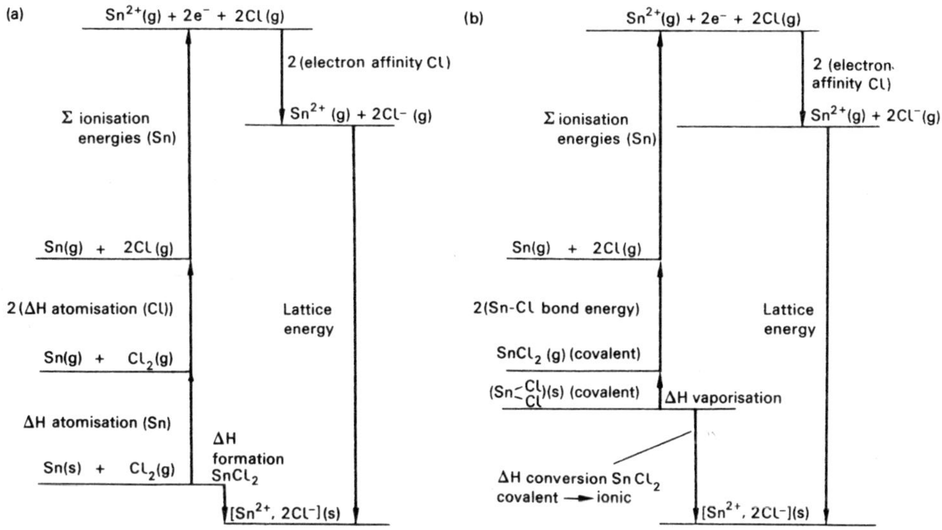

Fig. 5.1 Stability of tin(II) chloride ($Sn^{2+}2Cl^-$)(s): (*a*) *with respect to the elements;* (*b*) *with respect to covalent bonding.*

In the case of lead, the divalent state is more stable than the tetravalent state. Lead tetrachloride decomposes to lead(II) chloride at about room temperature, and lead(IV) oxide to lead(II) oxide on heating. Lead(IV) oxide is a good oxidising agent, converting hydrochloric acid to chlorine and hydrogen sulphide to sulphur with spontaneous ignition.

$$PbO_2 + 4HCl \longrightarrow PbCl_2 + Cl_2 + 2H_2O$$
$$PbO_2 + 2H_2S \longrightarrow PbS + S + 2H_2O$$

In the case of tin compounds, the divalent state is less stable than the tetravalent state. Tin(II) chloride is a good reducing agent in solution, converting mercury(II) ions to mercury, and iodine to iodide ions. Finely divided tin(II) oxide catches fire in air to form tin(IV) oxide.

$$\left\{\begin{array}{ll} Sn^{2+} + 2Hg^{2+} \longrightarrow Sn^{4+} + 2Hg^+ & \text{(white ppt. of HgCl in presence of Cl}^-\text{)} \\ Sn^{2+} + 2Hg^+ \longrightarrow Sn^{4+} + 2Hg & \text{(black ppt.)} \end{array}\right.$$
$$Sn^{2+} + I_2 \longrightarrow Sn^{4+} + 2I^-$$

Germanium forms fewer and less stable divalent compounds, principally the iodide and sulphide, whereas silicon forms no such stable compounds at room temperature. Silicon monoxide has been reported at 2000 °C. The reactions of the dichlorides with chlorine provides a striking comparison of stability.

CCl_2 and $SiCl_2$ occur only transiently at room temperature
$GeCl_2 \longrightarrow GeCl_4$ violent and uncontrollable
$SnCl_2 \longrightarrow SnCl_4$ slow at room temperature

but: $PbCl_4 \longrightarrow PbCl_2$ when just warm.

Unfilled orbitals are a feature of electronic structure just as much as filled ones! In particular, the outer valence shell of carbon (principal quantum number 2) can only accommodate eight electrons, whereas that of silicon can accommodate eighteen. Silicon (and other members of the group) can form six-co-ordinated complexes. Each bond contains two electrons, giving a total of twelve, so that the maximum is not exceeded. The formation of one such complex is responsible for the etching of glass by hydrofluoric acid.

$$SiO_2 + 6HF(aq) \longrightarrow \underset{\text{hexafluorosilicate(IV)}}{SiF_6^{2-}} + 2H^+(aq) + 2H_2O$$

Carbon, on the other hand, forms no such complexes. Thus, the hydrolysis of carbon tetrachloride cannot proceed via complex formation, and instead a hydroxide ion must displace a chloride ion directly.

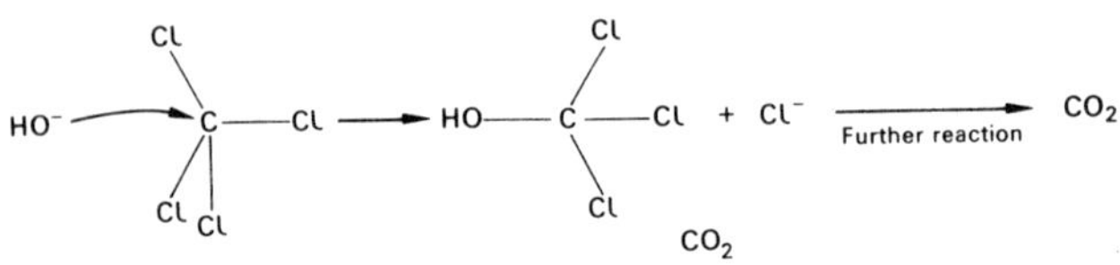

However, silicon tetrachloride can accept a hydroxide ion to form a low-energy intermediate complex:

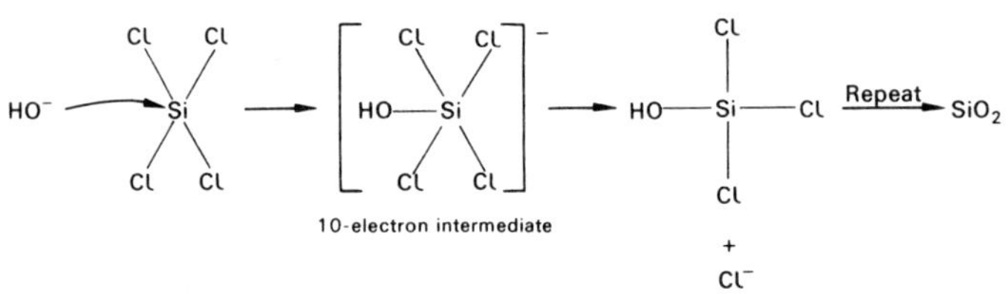

Hence, carbon tetrachloride is stable in contact with hydroxide solutions, but silicon tetrachloride is hydrolysed violently, even by water, to form hydrated silica.

$$SiCl_4 + 4H_2O \longrightarrow SiO_2.2H_2O + 4HCl$$

5.3 PROPERTIES RELATED TO ELECTRONIC STRUCTURE AND SIZE

5.3.1 Metallic character

'Metallic character', examined in detail in section 2.2.8, is a complex of features which are related to the attraction of a nucleus for the valence electrons. Large atoms hold electrons weakly, to form positive ions if the electrons are lost or very weak covalent bonds if they are shared. The boundary between metals and non-metals in group IV occurs at germanium, which can be considered as either, depending on the feature concerned.

For example, take the structure of the element itself. Carbon forms strong covalent bonds, and if these are tetrahedrally distributed, it takes the crystalline form of diamond. (The other form, or allotrope, is graphite, which is peculiar to carbon, and is considered later in the chapter.) Silicon and germanium crystallise with the same structure as diamond. They are all very hard, because of the three-dimensional interlocking structure, with diamond the hardest on account of the highest bond energy. Lead, however, forms the twelve-coordinated, close-packed metallic structure (slightly distorted), with its outermost electrons very mobile, thus contributing to the high conductivity of the metal. It is also typically malleable and ductile.

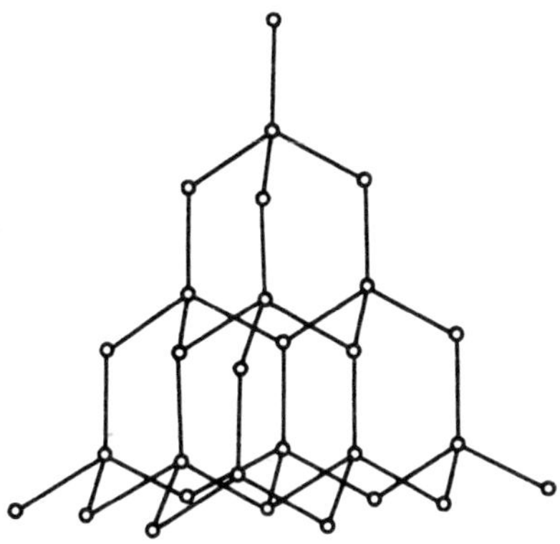

Fig. 5.2 Diamond crystal structure.

Tin, for once, occupies the intermediate position, having three allotropes. The low-temperature allotrope, α- (grey) tin, has the diamond structure, whereas the β- (white) and δ-tin have distorted close-packed structures. The close-packing results in a higher density than the α-form (7.3 $g cm^{-3}$ as opposed to 5.7 $g cm^{-3}$) so that, on prolonged cooling, the change to α-tin causes a disastrous expansion and powdering, known as tin plague.

Reaction with acids and alkalis also distinguishes metals and non-metals. Metals are more likely to dissolve in acids, forming cations, but non-metals dissolve to form covalently-bound oxo-anions. Silicon, for instance, dissolves in hot concentrated sodium hydroxide to form silicates.

$$Si + 2OH^- + H_2O \longrightarrow SiO_3^{2-} + 2H_2$$

Germanium and tin react similarly but more slowly. Lead does not react in this way but lead, on the other hand, dissolves in dilute nitric acid on heating to form lead(II) ions.

$$Pb + HNO_3 \longrightarrow Pb^{2+} + H_2O + NO/NO_2 \quad \text{etc.}$$

Tin will also dissolve in dilute acid, but germanium only with difficulty in concentrated acid, to form the dioxide.

5.3.2. Electronegativity

The positive nucleus of a small atom is able to attract and influence the electrons in the bonds that it forms, thus making the bond 'polar', or unsymmetrical in charge. This effect may extend to bonds further out, e.g. in the system X–O–H, where electrons are withdrawn from the O–H bond, allowing the hydrogen to be 'picked up' by a base (see sections 4.1 and 4.2). Thus carbonic and silicic acids, $CO(OH)_2$ and $Si(OH)_4$, form salts with bases. These acids are unstable and dehydrate readily, so that in practice the oxides CO_2 and SiO_2 are used.

$$\underset{}{CO_2} + \underset{\text{aq. solution}}{NaOH} \longrightarrow \underset{\text{hydrogencarbonate}}{NaHCO_3} \xrightarrow[-H_2O]{+CO_3} \underset{\text{carbonate}}{Na_2CO_3}$$

$$SiO_2 + \underset{\text{fused}}{NaOH} \longrightarrow \underset{\text{silicate}}{Na_2SiO_3} + H_2O$$

This difference in reactivity is a kinetic effect, due to the high activation energy involved in breaking the strong Si–O bond. The acidity of each oxide is comparable, and this is confirmed by the following equilibrium:

$$SiO_2 + Na_2CO_3 \rightleftharpoons Na_2SiO_3 + CO_2$$

Forward reaction favoured by high temperature, owing to loss of CO_2 from system as gas (Le Chatelier).
Reverse reaction favoured by aqueous conditions, owing to loss of silicon(IV) oxide from system as precipitate.

Germanium(IV) oxide is less acidic, and will even, with difficulty, dissolve in concentrated acids thereby showing some amphoteric character. Tin(IV) and lead(IV) oxides are definitely amphoteric, dissolving freely in concentrated acid or alkali.

$$XO_2 + 4HCl \longrightarrow XCl_4 + 2H_2O \quad \text{(cold concentrated acid)}$$
$$2H_2O + XO_2 + 2NaOH \longrightarrow Na_2X(OH)_6 \quad \text{(hot aqueous alkali)}$$

Notice the stability of the O–H bond in the hexahydroxoplumbate(IV), $Pb(OH)_6^{2-}$, compared with its absence in carbonates, CO_3^{2-}, due to the polarising power of small atoms referred to earlier (see sections 4.1 and 11.3.1).

Tin and lead form divalent oxides and hydroxides, and in these the metal atom carries a formal positive charge of +2. The attraction for electrons in neighbouring bonds decreases compared to the +4 state, and so the hydroxides (and oxides) dissolve more readily in acid than in alkali.

$(HO)_3\overset{(4+)}{Pb}—O\overset{\downarrow}{—}H$	$HO\overset{(2+)}{Pb}\overset{\downarrow}{—}O–H$
strong attraction for electrons H^+ readily lost acidic oxide or hydroxide	weak attraction for electrons OH^- readily lost basic oxide or hydroxide

5.4 PROPERTIES RELATED TO BOND ENERGIES

5.4.1 **Catenation (the formation of chain molecules)**

The estimated values of the X–X bond enthalpies in $kJmol^{-1}$ are

C–C 350, Si–Si 220, Ge–Ge 160, Sn–Sn 150, Pb–Pb 120.

Notice the high value for C–C compared to other members of the group. In contrast, all values for X–Cl lie between 340 and 380 $kJmol^{-1}$. Compounds containing chains of carbon atoms are stable for two reasons. Firstly, if the enthalpy of a reaction such as

$$Cl_2 + {>}C—C{<} \longrightarrow 2\,{>}C–Cl$$

is calculated, it is typically 100 to 200 kJ less exothermic than a similar reaction involving the other group IV elements, due to the extra energy needed to break the C–C bond. Secondly, and more important, the increased energy of activation of the step in which the C–C bond is broken will be enough to cause a drastic reduction in the rate of reaction.

In practice, this results in a marked increase in the number of compounds containing carbon chains. About three million of these are known (and studied in organic chemistry), containing as many as twenty or thirty carbon atoms in a chain, and with chlorine, oxygen and many other atoms as substituents. Catenation is much less common in silicon and germanium compounds, although hydrides up to Si_6H_{14} and Ge_5H_{12} are known, together with some halides. Tin and lead form no chain compounds. Significantly the catenated compounds of silicon and germanium are very reactive and may ignite spontaneously in air

$$Si_2H_6 + 3\tfrac{1}{2}O_2 \longrightarrow 2SiO_2 + 3H_2O$$

or be hydrolysed by acid or base.

5.4.2 **Hydrides, XH_4**

The bond energy of the hydrides decreases from carbon to lead. The values in $kJmol^{-1}$ are C–H, 410; Si–H, 320; Ge–H, 320; Sn–H, 310; Pb–H, 290. In general, the hydrides can be prepared by the action of acid on a carbide, silicide, etc., and are all gaseous, being simple covalent molecules.

$$Mg_2Si + 4H^+ \longrightarrow 2Mg^{2+} + SiH_4$$

The stability decreases from carbon to lead. Methane, CH_4, is stable in air until ignited, but all the others are spontaneously flammable. Stannane, SnH_4, and plumbane, PbH_4, decompose spontaneously at room temperature.

5.4.3 **Structure of oxides**

Only for carbon is the X=O bond more than twice as strong as the X–O bond. The values are 743 $kJmol^{-1}$ and 360 $kJmol^{-1}$ respectively, so that the enthalpy of the reaction in fig. 5.3 can be found.

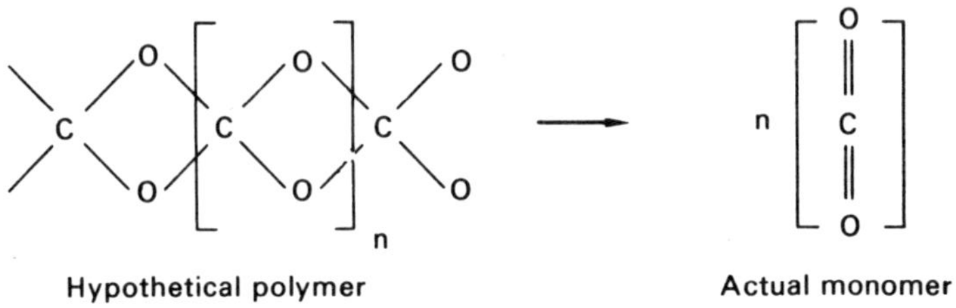

Fig. 5.3 Alternative structures for carbon dioxide.

For each CO_2 unit converted, four C–O bonds are broken (two within the square bracket, and a 'half-share' of the four bonds joining two units) and two C=O bonds are formed. Net enthalpy change equals $(4 \times 360) - (2 \times 743) = -46\ kJmol^{-1}$.

The value of $-46\ kJmol^{-1}$ indicates that the reaction goes to the right, and so carbon dioxide exists as single covalent molecules making up a gas. However, the situation is reversed for silicon(IV) oxide (silica). There are many crystalline forms of silica based on the single-bonded unit, in which corners may be shared to form a three-dimensional structure of high melting point as in quartz.

Alternatively some corners may be shared and some paired with cations as in the inorganic polymer glass.

Tin and lead(IV) oxides have a structure in which six oxide ions are symmetrically arranged about each metal ion. Such a symmetrical crystal is typical of many ionic substances.

5.5 ANOMALIES DUE TO π-BONDING IN CARBON

Carbon is the only element in group IV which forms strong π-bonds. If, for example, a carbon atom forms bonds with three neighbouring atoms only the first three pairs of electrons can form bonds along the line of centres of the nuclei (fig. 5.5, I). The fourth pair occupies a site so as to minimise the repulsion between it and the existing electrons and to maximise the attraction between it and the positive nuclei. The bond formed is the π-bond and takes up the position shown in (II). Because the π-bond is off the line of centres, it becomes rapidly weaker as the size of the atom increases.

Graphite is an allotrope peculiar to carbon, which relies for its stability on the presence of an extended network of π-bonds associated with each plane of atoms.

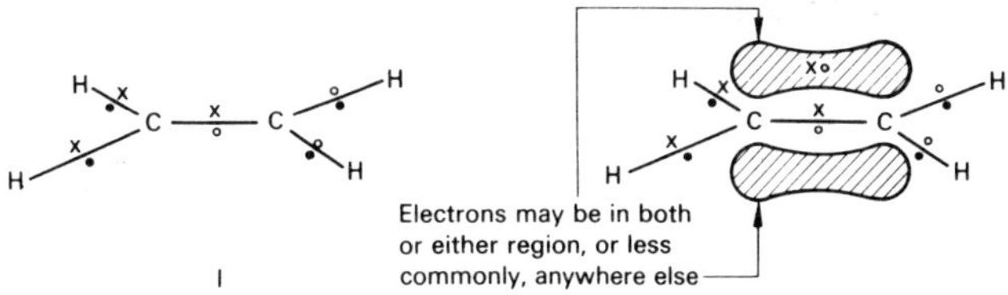

I II

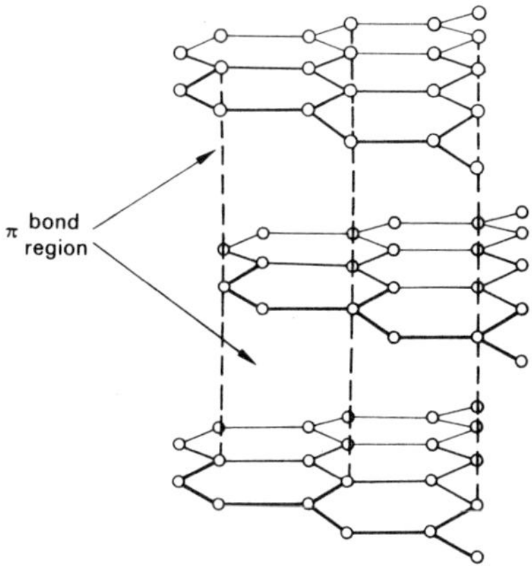

III Graphite

Fig. 5.4 π bonding in carbon chemistry.

The π-electrons are weakly held, and will move in an electric field, so that graphite conducts electricity. Furthermore, bonding between the planes is very weak, so that sliding can occur, and graphite is indeed widely used as a lubricant.

Carbon monoxide is also exceptional. Its bond length indicates that it does not exist as the six-electron compound (fig. 5.6, I), but that a second π-bond is formed using non-bonding electrons on the oxygen atom (II). The extra bonding shortens the C–O distance.

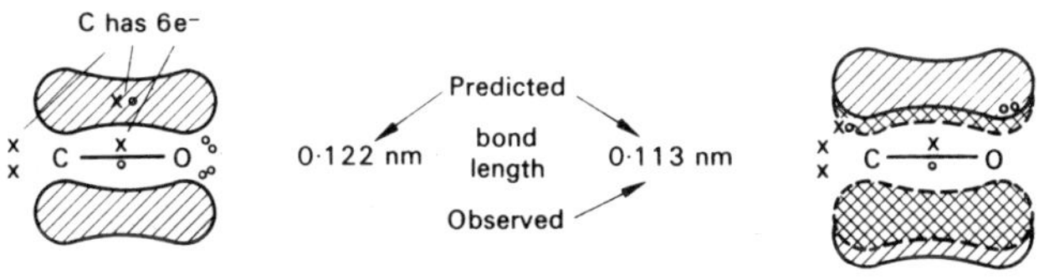

I II

Fig. 5.5 Structure of carbon monoxide.

The abundance of weakly held π-electrons renders carbon monoxide reactive. It is readily oxidised to tetravalent carbon:

$$CO + \tfrac{1}{2}O_2 \longrightarrow \underset{\text{carbon dioxide}}{CO_2} \quad \text{(on ignition)}$$

$$CO + Cl_2 \longrightarrow \underset{\text{carbonyl chloride}}{COCl_2} \quad \text{(on heating)}$$

$$CO + PbO \longrightarrow CO_2 + Pb \quad \text{(on heating)}$$

5.6 SUMMARY OF GENERAL PROPERTIES

	C	Si
State of element at room temperature and pressure	(1) Diamond. Solid, giant molecule, very hard, transparent (2) Graphite. Solid, giant molecule, soft, grey	Solid with diamond structure, hard, brittle, black and shiny
Electronic structure	2:4	2:8:4
Radius of atom/(nm)	0.077	0.117
First four ionisation energies/($MJmol^{-1}$)	1.1, 2.3, 4.6, 6.2	0.8, 1.6, 3.2, 4.4
Electrical conductivity ($\mu\Omega^{-1}cm^{-1}$)	negligible (diamond)	1.0
Atomic mass and mass of common isotopes	12.0 (12, 13)	28.1 (28, 29, 30)
Density/ ($kg\ m^{-3} \times 10^{-3}$)	2.25 (gr) 3.5 (dia)	2.3
Melting point/(°C)	3700	1400
ΔH *atomisation*/ ($kJmol^{-1}$)	715	440
Boiling point/(°C)	4800	2400
Occurrence	Organic matter, carbon dioxide in atmosphere and oceans Carbonate rocks ($CaCO_3$, limestone)	Silicates, involving K Na Ca Al Fe Mg, in many rocks Sand (mainly SiO_2)
Extraction	Partial oxidation of organic material in limited supply of air Oil $\longrightarrow$ lampblack Wood, bones $\longrightarrow$ charcoal	Sand and coke in electric furnace $SiO_2 + 2C \longrightarrow Si + 2CO$ At higher temperatures, an excess of coke is converted to graphite

and forms complexes with many metals,

$$\text{e.g.}\quad Ni + 4CO \xrightarrow{100°C} \underset{\text{nickel carbonyl}}{Ni(CO)_4}$$

The strength of π-bonding in carbon compounds also accounts for the gaseous nature of carbon dioxide (see above).

AND REACTIONS OF GROUP IV

Ge	Sn	Pb
As silicon	Silvery metal, tough, malleable, conductive (Low temperature, powdery allotrope formed)	Dull metal, soft, ductile, conductive
2:8:18:4	2:8:18:18:4	2:8:18:32:18:4
0.122	0.141	0.154
0.8, 1.5, 3.1, 4.4	0.7, 1.4, 2.9, 3.9	0.7, 1.4, 3.0, 4.0
2.0	8.0	5.0
72.6 (70, 72, 74)	118.7 (116, 118, 120)	207.1 (206, 207, 208)
5.3	5.7 (α) 7.3 (β)	11.3
940	230	330
380	300	200
2800	2200	1700
As a complex sulphide	Cassiterite SnO_2 S.E. Asia W. Africa Bolivia (a little in Cornwall)	Galena PbS Canada Australia (a little in the Pennines)
Extract with acid as $GeCl_4$, hydrolyse to GeO_2, and reduce with hydrogen	Roast ore with coke $SnO_2 + 2C \longrightarrow Sn + 2CO$	Roast ore in air and then with coke $PbS + \frac{3}{2}O_2 \longrightarrow PbO + SO_2$ $PbO + C \longrightarrow Pb + CO$

	C	Si
Uses of elements	Gemstones, drills, saws (diamond) Lubricants, pencils (graphite) Refining and decolorising sugar or pharmaceuticals (porous charcoal); rubber filler and inks (lampblack)	Alloyed with iron as a reducing agent for MgO in industrial extraction of Mg
Uses of compounds	Carbon dioxide for audification (e.g. aspirin production) for refrigeration as dry ice, in fizzy drinks and extinguishers Fluorocarbons (e.g. CF_2Cl_2) as refrigerants and aerosols See also p. 30 for $CaCO_3$	Glass is a mixture of silica and sodium silicate. Sodium silicate is used as adhesive in cardboard boxes Silica is also used for high temperature refractory crucibles
Reactions of element $X + O_2 \longrightarrow XO_2$	Burns at red heat	Burns at red heat
$X + 2OH^- + H_2O \rightarrow XO_3^{2-} + H_2$	No reaction	In hot concentrated alkali
$X + 4HNO_3 \longrightarrow XO_2 + 2H_2O + 4NO_2$	Dissolves in fuming nitric acid, heated	No reaction
Other reactions	Reacts with metal oxides to form metals (Pb, Cu, etc.)	Concentrated HF $\longrightarrow$ H_2SiF_6: hot $Cl_2 \longrightarrow SiCl_4$: hot Mg $\longrightarrow Mg_2Si$
Preparation of XO_2	Dilute acid + carbonate $2H^+ + CO_3^{2-} \longrightarrow CO_2 + H_2O$ Collect by downward delivery	Occurs naturally or by hydrolysing $SiCl_4$ and heating product
$XO_2 + 2OH^- \longrightarrow XO_3^{2-} + H_2O$	Rapid with dilute alkali	Slow with fused alkali
$XCl_4 + 2H_2O \longrightarrow XO_2 + 4HCl$	No reaction	Rapid with cold water

Ge	Sn	Pb
In transistors and rectifiers	Tinplate (canning) accounts for 50%. Minor uses include bearing alloys, bronze and solders	Batteries, radiation shielding, and, decreasingly, in the building trade
–	Tin(IV) oxide for making opaque glass Organo-tin compounds as pesticides.	Lead tetraethyl as anti-knock in petrol Various pigments (e.g. $PbCO_3$, $PbCrO_4$) and glazes (PbO) Stabilisers in polyvinyl chloride ($PbCO_3$)
Slow at red heat	Slow reaction at white heat	PbO formed at 400 °C
In hot concentrated alkali	Slow in hot concentrated alkali	No reaction
In hot concentrated acid	In hot concentrated acid	$Pb(NO_3)_2$ formed in cold concentrated acid
Hot $Cl_2 \longrightarrow GeCl_4$	Concentrated $HCl \longrightarrow Sn^{2+}(aq)$ hot $Cl_2 \longrightarrow SnCl_4$	hot $Cl_2 \longrightarrow PbCl_2$
By hydrolysing $GeCl_4$ and heating product	By heating metal with con-concentrated HNO_3	Dilute acid on red lead (Pb_3O_4)
Slow with fused alkali	Dissolves in concentrated alkali	Dissolves in concentrated alkali
In cold water	Reversible in cold water	Reversible and slow

5.7 SPECIAL REACTIONS OF CARBON COMPOUNDS

5.7.1 **Carbon monoxide**

(colourless, odourless and intensely poisonous; insoluble in water).

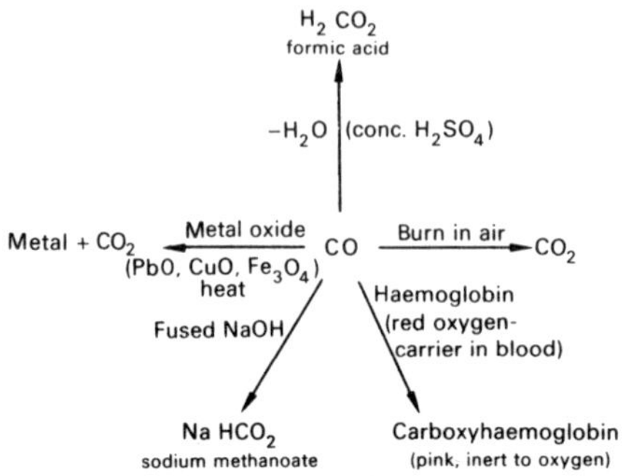

Fig. 5.6 Reactions of carbon monoxide.

5.7.2 **Carbon dioxide in the biosphere**

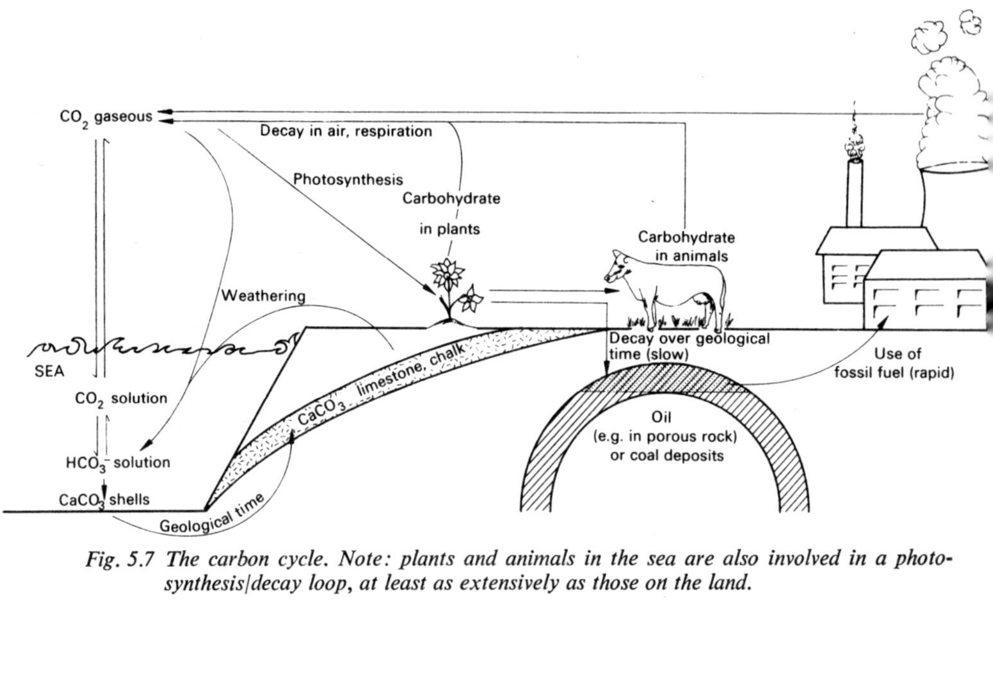

Fig. 5.7 The carbon cycle. Note: plants and animals in the sea are also involved in a photosynthesis/decay loop, at least as extensively as those on the land.

5.8 REACTIONS OF LEAD AND TIN

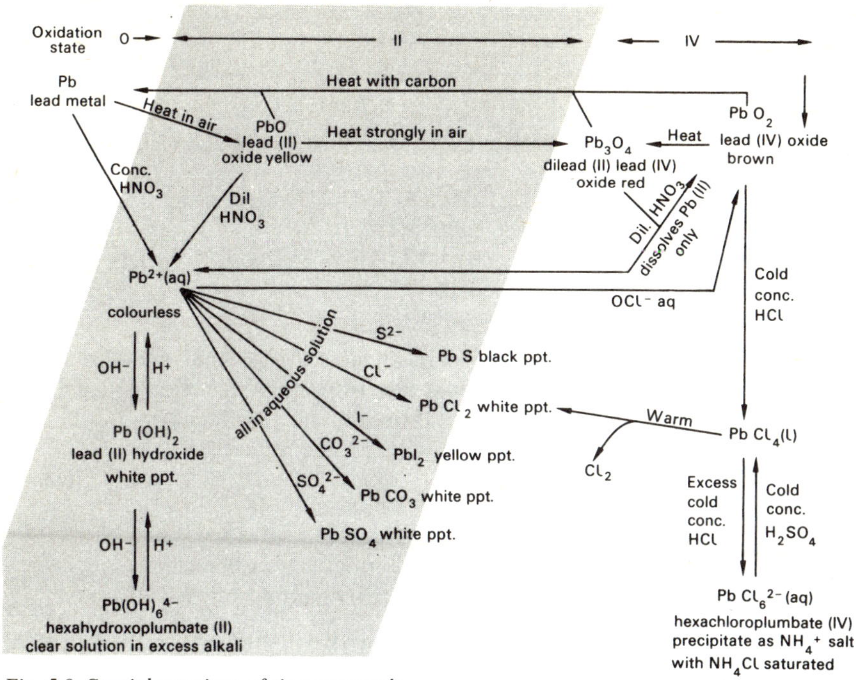

Fig. 5.8 Special reactions of tin compounds.

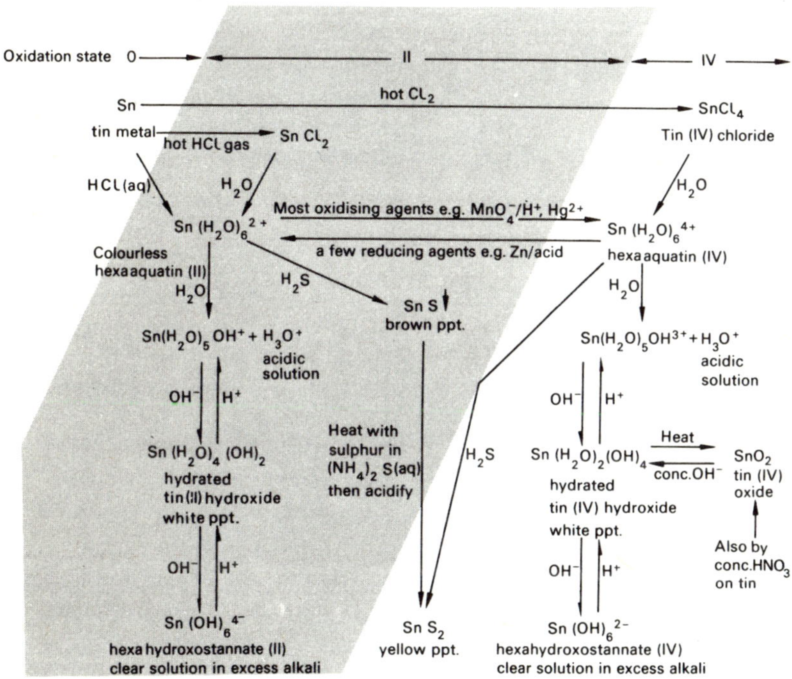

Fig. 5.9 Special reactions of lead compounds.

QUESTIONS

1 Show how the following properties vary from carbon to lead. Quote observable phenomena in illustration, and relate the trend to the electronic structures and sizes of the atoms.
(a) The relative stability of the +IV and +II oxidation states;
(b) The crystal structures of the elements;
(c) The acidity/basicity of the oxides XO_2;
(d) The thermal stability of the hydrides;
(e) One other property of your choice.

2 What features of the carbon atom are responsible for the following exceptional properties? Show how the feature is responsible for the property.
(a) CCl_4 does not react with water: all the other tetrahalides do.
(b) One allotrope of carbon is a good electrical conductor.
(c) There are more compounds of carbon known than of all other elements put together.
(d) Carbon monoxide is a gas stable at room temperature.
(e) Carbon has a higher melting point than any other element.

3 Describe the carbon cycle in good prose which could be understood by someone without technical training. Go on to discuss the implications for energy policies based on (a) fossil fuel, (b) annual wood crops, (c) nuclear power. (In 1975 the annual world consumption was approximately (a) 5×10^9, (b) 10^6, (c) 10^7 tons of coal equivalent.)

4 Relate each observation given below to its corresponding principle.

Observation	*Principle*
(a) All known life is based on compounds of carbon.	There are many stable compounds with long chains of carbon atoms.
(b) Glasgow residents (in older property) are advised not to drink water which has stood in pipes overnight.	Lead has a negative redox potential.
(c) Hydrofluoric acid is used to etch glass.	Silicon can form complexes by expanding its outer shell to a maximum of eighteen electrons.
(d) Sandstone is part of the lithosphere (rocks) but carbon dioxide is part of the atmosphere.	Si–O bonds are very strong compared to Si=O bonds.

5 There is evidence to the effect that of all atmospheric pollutants, lead is found in the human body in concentration closest to that known to cause clinical symptoms of poisoning. If it were decided to ban the use of lead and its compounds, what products would be affected? Suggest possible substitutes. Can you justify not using these substitutes?

6 Identify the following elements and compounds, account for the data given and write equations for the changes occurring.

(a) An element A has a relative atomic mass of 118.7 but when sampled in a mass spectrometer shows peaks at 116, 118 and 120.

(b) An element B reacts with very nearly twice its mass of chlorine on heating to form a liquid. This liquid reacts with water to form a white solid, which, when heated with hydrogen re-forms the element.

(c) A white solid C when heated loses about one-sixth of its mass and becomes deep yellow. Further heating in air results in a gain in mass and the colour turns to red. This solid partly dissolves in dilute nitric acid leaving a brown residue, but the original solid dissolves completely with effervescence to yield a clear solution. The original solid does not dissolve appreciably in dilute sulphuric acid.

7 Explain the following observations:

(a) Tests involving lead cations in solution always specify the use of nitrate(V) or acetate.

(b) The proportion of sodium, potassium and calcium ions in glass can be varied over a wide range without much altering the properties of the glass.

(c) Carbon monoxide is a more dangerous gas than hydrogen sulphide.

(d) Solutions of tin(IV) compounds are more acidic than solutions of tin(II) compounds.

(e) Molten $PbCl_2$ conducts electricity better than molten $PbCl_4$.

8 The following statements are not correct in all respects. There may be exceptions, ambiguities, omissions or deliberate mistakes. Write out a correct version of the statement and comment on the theoretical background to it. All statements refer to group IV elements.

(a) The elements are prepared by the reduction of XO_2 with carbon.

(b) The elements dissolve with reaction in fused sodium hydroxide.

(c) All the tetrahydrides are unstable in the presence of air.

(d) The tetrahalides are formed on heating the dihalides in chlorine.

9 Write balanced equations for the following reactions:

(a) Lead(II) oxide $\xrightarrow{\text{dilute nitric (V) acid}}$ lead(II) nitrate.

(b) Germanium $\xrightarrow[\text{hydroxide}]{\text{sodium}}$ sodium trioxogermanate(IV).

(c) Silicon $\xrightarrow{\text{hydrofluoric acid}}$ fluorosilicic(IV) acid.

(d) Hexaaquatin(IV) $\xrightarrow[\text{hydroxide}]{\text{dilute sodium}}$ hydrated tin(IV) hydroxide.

(e) Formic acid $\longrightarrow$ carbon monoxide.

(f) Carbon $\xrightarrow[\text{acid}]{\text{concentrated nitric}}$ carbon dioxide

6

Group V: Nitrogen, Phosphorus, Arsenic, Antimony, Bismuth

A contrasting types of set of bonds

6.1 INTRODUCTION

The elements of group V—nitrogen, phosphorus, arsenic, antimony and bismuth—share the common electronic structure in which the pattern is always ns^2np^3. However, this allows a wide variation in the types of compounds formed, rather than imposing common features upon the elements in the group. An element with five electrons can form bonds in the following ways:

Arrangement of electrons	*Description of bond*	*Examples and comments*
(1) An extra three electrons are gained, to form an octet	Ionic, as anion X^{3-} 	N^{3-} only. A large amount of energy is required to place the second and third electron on an ion which is already negative. Only where the electrons are close to the nucleus, as in the small N atom, is this possible. Compounds such as Na_3P, and Na_3As have crystal structures typical of metallic alloys
(2) Three electron pair bonds are formed, and again an octet is reached	Covalent 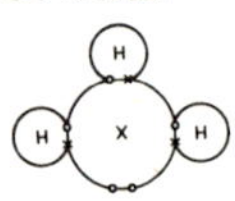	Many examples: NH_3, PH_3, AsH_3, SbH_3 and also the trihalides. Bismuth rarely forms bonds of this type, since, being large, it attracts its outer electrons weakly. Rather than take on three more, bismuth loses three to form a tripositive ion. (See (5))
(3) More than three electron pair bonds are formed, so that the outer shell is expanded beyond an octet	Covalent 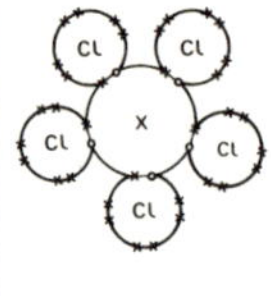	The pentahalides PBr_5, AsF_5, $SbCl_5$, etc. This type of bonding is not available to nitrogen, as the second principal shell cannot expand to ten or more electrons. Compounds such as NH_4Cl are in fact NH_4^+, Cl^-, with only four bonds direct to nitrogen. Occasionally more than ten electrons occur in the outer shell, the extra pairs being donated as a co-ordinate bond. Thus solid phosphorus pentachloride exists as $(PCl_4^+), (PCl_6^-)$ or $(\overset{\times\times}{\underset{\times\times}{{}^{\times}_{\times}Cl}} \xrightarrow[+]{\times} PCl_5)^-$

(4) The non-bonding pair of electrons in (ii) may be donated to an electrophile	Co-ordinate 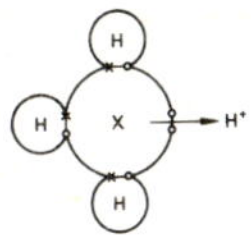	Ammonium salts (NH_4^+) and phosphonium salts (PH_4^+), in which a proton is co-ordinated. This type of bond is only common in nitrogen and phosphorus chemistry, but may be formed with other electrophiles besides the proton e.g., the electron deficient boron halides 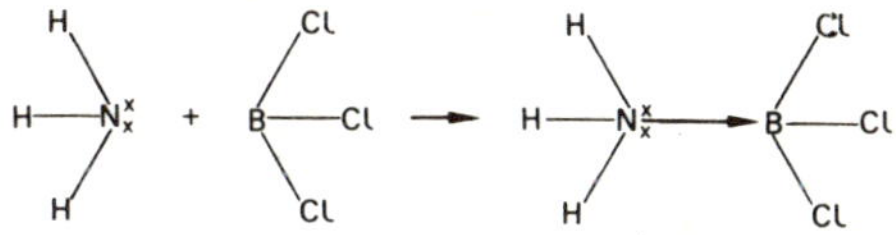
(5) Electrons may be lost to form cations	Ionic, as cation X^{3+}	No X^{5+} ions are known, nor would they be predicted from the cumulative amount of energy required to cause this fivefold ionisation. Cations are most likely to be found in elements whose electrons are furthest from, and weakly held by, the positive nucleus. Thus, Bi^{3+} occurs in solids and, as the hexaaqua ion, in solution, and the stability of these two ions is attributed to the 'inert pair effect'—see chapter 4
(6) Other cases		Nitrogen also forms hydrogen bonds and electron-deficient (i.e. with less than an octet) covalent molecules

Table 6.1 Bonding in group V.

6.2 SPECIAL FEATURES OF NITROGEN CHEMISTRY WHICH CONTRAST WITH THE REST OF THE GROUP

6.2.1. Bond energies

The triple bond energy for nitrogen is quite exceptional in comparison with its neighbours in the periodic table. The values in $kJmol^{-1}$ are given in table 6.2.

C—C 350	C=C 600	C≡C 810
N—N 160	N=N 420	N≡N 950
P—P 210	(P=P not known)	P≡P 490

Table 6.2 Bond energies for nitrogen and related elements.

The N ≡ N bond is about six times as strong as the N–N single bond, unlike both carbon and phosphorus, for which the corresponding ratio is about $2\frac{1}{2}$. Furthermore, the first stage of the breakdown of the triple bond, to give X=X, requires about 500 $kJmol^{-1}$, as against about 200 $kJmol^{-1}$ in the other two cases. We can therefore predict three things.

Firstly, nitrogen will exist as the diatomic gas, whereas carbon and phosphorus will have allotropes in which single bonds are dominant. Phosphorus vapour, in fact, exists as P_4 molecules which are tetrahedra of phosphorus atoms. On condensing the vapour rapidly, these persist, forming the allotrope, yellow phosphorus. However, this is unstable and changes slowly to red phosphorus, which is believed to have either a chain or a sheet structure.

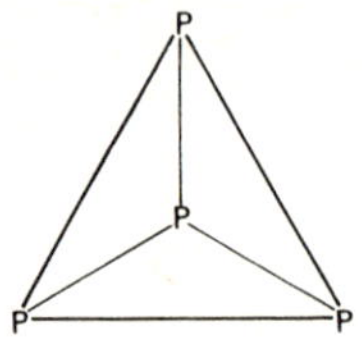

Fig. 6.1 The yellow phosphorus molecule.

Secondly, because of the energy of formation of the triple bond, many nitrogen compounds will be endothermic (formed from their elements with absorption of energy) and correspondingly unstable. Nitrogen probably forms more types of endothermic compounds than any other element. There are, for example, oxides (from NO, $\Delta H_f = 90$ kJmol^{-1} to N_2O_4, $\Delta H_f = 10$ kJmol^{-1}), hydrides (N_2H_4, $\Delta H_f = 50$ kJmol^{-1}) and halides (NCl_3, $\Delta H_f = 40$ kJmol^{-1}). This feature of nitrogen chemistry explains its use in so many explosives, from azides, N_3^-, used as detonators, to trinitrotoluene (TNT), $C_7H_5N_3O_6$, or nitroglycerine (dynamite), $C_3H_5N_3O_9$.

Thirdly, the large energy difference between the triple and double bond is associated with a high activation energy for many reactions. Hence, nitrogen is very unreactive. It acts as a diluent to its active partner in the atmosphere, oxygen. It reacts at room temperature only with lithium, to form the nitride, and with the nitrogen-fixing enzyme systems of bacteria. At higher temperatures it will react with hydrogen—perhaps the chemical reaction most essential to our standard of living—as described on page 71; and also with magnesium or oxygen.

6.2.2 Size of atom

The covalent radius of the nitrogen atom is about three-quarters that of the phosphorus atom. This is typical of elements at the head of a group; the ratio of the radii of head element/second element is often smaller than the ratio second element/largest element.

In this case there is a direct consequence because the nitrogen atom is small enough to fit in the holes left in the close-packed lattice of most metal crystals. Such 'interstitial compounds' are discussed more fully under hydrogen. Nitrogen, being a larger atom than hydrogen, cannot escape from the metal lattice as readily, and forms a stable 'compound'. The foreign atoms act as pegs to stop the layers of the metal crystal sliding, and the melting point and hardness increase sharply. Practical use of this is found in case-hardening, a process in which steel is plunged into a canister of sodium cyanide pellets, forming interstitial carbides and nitrides on the surface. Gun-barrels are treated in this way. Besides iron, the transition elements from Sc to Cr also form such nitrides readily.

6.2.3. Formation of co-ordinate bonds

The exceptional formation of a nitride ion, N^{3-}, as a result of the small size of the nitrogen atom has already been dealt with, but there are other effects upon bonding related to size. Ammonia, for instance, is decidedly basic, whereas phosphine is very weakly basic. The electron density due to the unpaired electrons in the ammonia molecule is much greater than that in phosphine, and therefore electrophiles, such as the proton, are more readily attracted away from competing bases. With strong acids (HCl, H_2SO_4, etc.), the transfer is complete, and ammonium chloride, for instance, is stable at room temperature. Ammonia will react reversibly with weaker acids at room temperature: carbonic acid forms solid ammonium carbonate, and in water the hydroxide ion is formed.

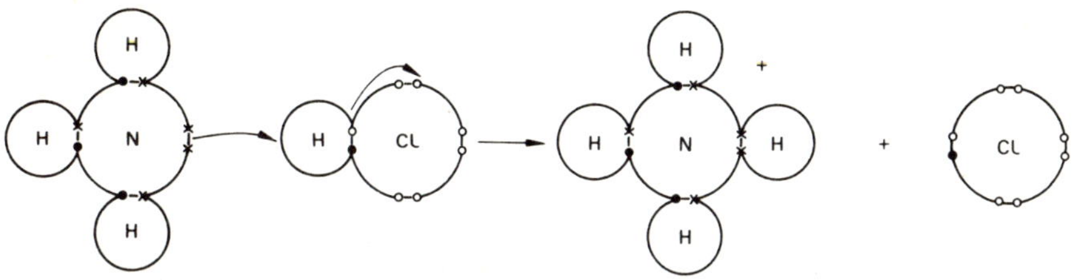

Fig. 6.2 Electron transfer in the formation of ammonium chloride.

$$NH_3 + H_2O + CO_2 \rightleftharpoons NH_4^+ + HCO_3^-$$
$$NH_3 + H_2O \rightleftharpoons NH_4^+ + OH^-$$

In both cases, there is still a strong smell of ammonia showing that the reaction is reversible.

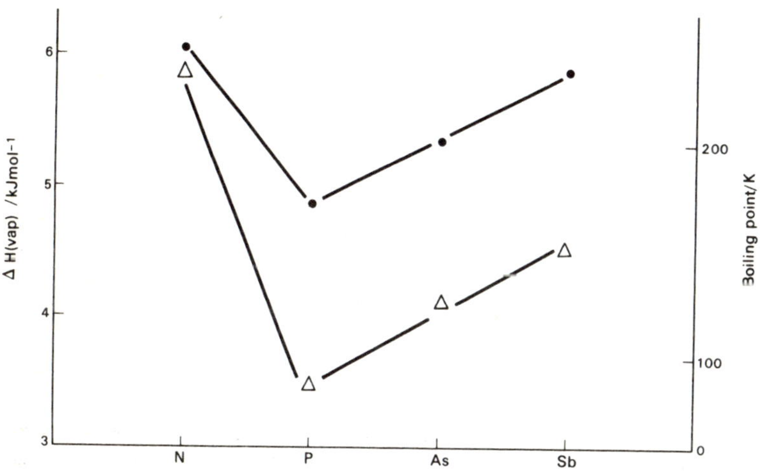

Fig. 6.3 Boiling points (·) and heat of vaporisation of XH_3 (Δ).

In contrast, the only phosphonium salt stable at room temperature is the iodide, $PH_4^+I^-$, all others dissociating into phosphine and the acid. Both ammonia and phosphine form a large variety of complexes with transition metals by donation of these non-bonding electrons to the metal ion.

The solubility of ammonia in water is due not to the above ionisation (in 1 M ammonia, only about 0.004 M is present as NH_4^+) but primarily to hydrogen bonding, these bonds being formed between ammonia and water molecules. Hydrogen bonding is characteristic of small electronegative atoms such as oxygen and nitrogen (see section 9.2.2). The boiling point and heat of vaporisation of ammonia, compared to the other hydrides, also indicate the presence of such bonds, holding one molecule to the next and hindering vaporisation.

6.2.4 **Oxides of nitrogen and phosphorus**

Some oxides of nitrogen owe their existence to the availability of the non-bonding electron pair on nitrogen. Dinitrogen monoxide (B) can be regarded as a molecule of nitrogen (A) which has formed a co-ordinate bond to an atom of oxygen; although the actual electron distribution will be an average of the structures (B) and (C), the latter having an 'internal' co-ordinate bond. The weakness of this bond is reflected in the easy decomposition of dinitrogen monoxide at low temperatures to nitrogen and oxygen. Most substances which burn in air generate enough heat to decompose dinitrogen monoxide and continue to burn in the atmosphere formed, which contains 30 per cent oxygen.

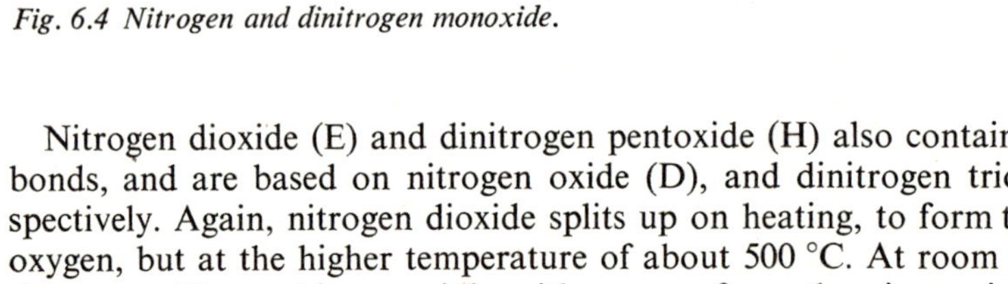

Fig. 6.4 Nitrogen and dinitrogen monoxide.

Nitrogen dioxide (E) and dinitrogen pentoxide (H) also contain co-ordinate bonds, and are based on nitrogen oxide (D), and dinitrogen trioxide (F) respectively. Again, nitrogen dioxide splits up on heating, to form the oxide and oxygen, but at the higher temperature of about 500 °C. At room temperature, the monoxide combines rapidly with oxygen from the air to give the brown dioxide. This is one of the rare termolecular gas reactions, the rate being proportional to $[NO]^2[O_2]$. The improbability required, that of getting three molecules together at once, explains the persistence of low concentrations of monoxide in the atmosphere derived from car exhausts in apparent contradiction of the rapid reaction cited above. It is a serious pollutant in some areas.

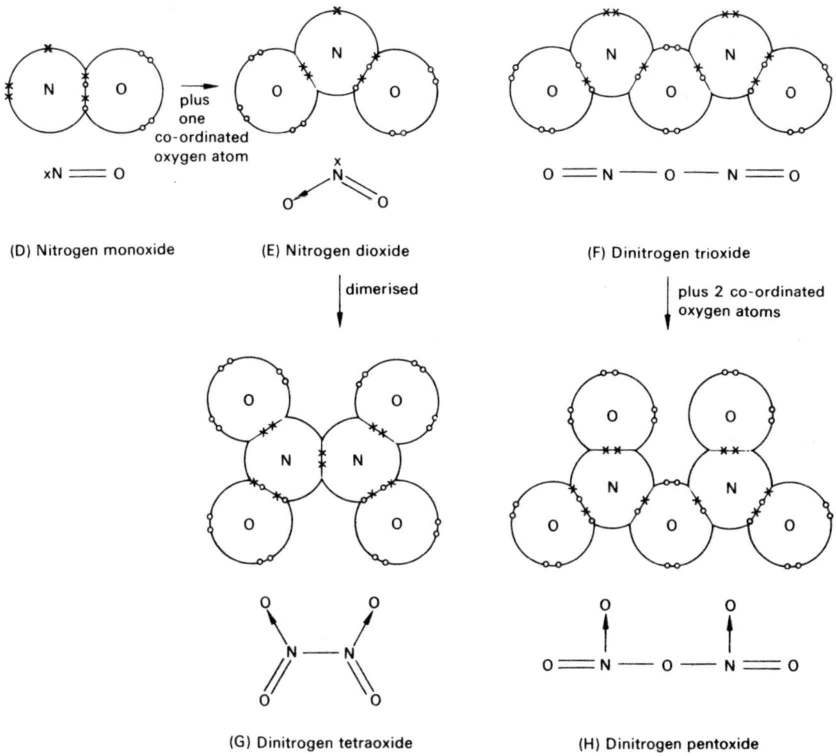

Fig. 6.5 Nitrogen monoxide and higher oxides of nitrogen.

The oxides of phosphorus are based on the P_4 tetrahedron found in the vapour phase. Phosphorus(III) oxide is in fact P_4O_6, oxygen atoms being inserted between each pair of phosphorus atoms, and phosphorus(V) oxide is P_4O_{10}, having additional oxygen atoms co-ordinated to each corner of the tetrahedron. There is, however, no evidence for such co-ordination to oxygen in any of the remaining oxides of the group.

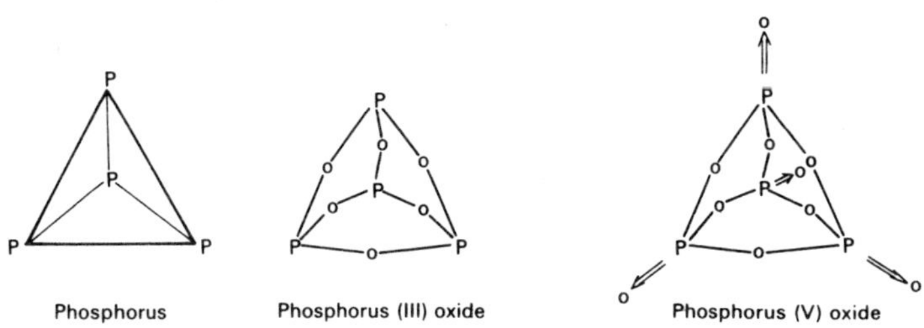

Fig. 6.6 Phosphorus and its oxides.

6.2.5 Expansion of outer shell beyond an octet

Such expansion is not possible for nitrogen. Hence nitrogen trichloride is known, but not the pentachloride. Both chlorides are known for phosphorus, but, not unexpectedly, the pentachloride dissociates, reversibly at room temperature and completely at higher temperatures, to yield the trichloride with its stable octet.

$$PCl_5 \rightleftharpoons PCl_3 + Cl_2$$

An associated trait, mentioned in chapter 4, is the tendency for small atoms to form fewer bonds. Thus nitric(V) and phosphoric(V) acid both have the same oxidation states, but the former has bonds to three atoms, and the latter to four atoms.

Nitric(V)acid Phosphoric(V) acid

Fig. 6.7 Nitric(V) and phosphoric(V) acids.

6.2.6 The number of oxidation states

Nitrogen exists in more oxidation states, and these are more commonly available, than any other member of the group. Table 6.3. shows representative members in states from +V to −III. In calculating relations for the equations shown in the table it is best to use the principle of balanced oxidation state changes. Thus 50 per cent nitric(V) acid is reduced to nitrogen monoxide by copper, which is itself simultaneously oxidised to copper(II) ions. The oxidation state change for nitrogen is (−III) and that for copper (+II). Since

$$HN^{V}O_3 \longrightarrow N^{II}O(\triangle = -III) \quad Cu \longrightarrow Cu^{2+}(\triangle = +II)$$

$$2\times(-III) + 3\times(+II) = O, \quad 2HNO_3 \equiv 3Cu$$

$$\left.\begin{array}{l} 3Cu + 2HNO_3 \longrightarrow 3Cu^{2+} + 2NO \\ 3Cu + 2HNO_3 + 6H^+ \longrightarrow 3Cu^{2+} + 2NO + 3H_2O \end{array}\right\} \text{make final correction using } H^+ \text{ ions and water}$$

6.2.7 Disproportion

Disproportion is the name given to the type of reaction in which a substance reacts with itself, part becoming oxidised and part becoming reduced. The products therefore contain the same element in two different oxidation states. For example:

$$N_2{}^{III}O_3 \longrightarrow N^{II}O + N^{IV}O_2$$

Several examples are to be found in the table on page 74. Disproportion reactions are often initiated by a change in the acidity of the environment. Thus, in alkaline conditions, nitrogen dioxide oxidises nitrogen oxide to nitrite(III)—a reaction used for the final purification of the waste gas stream in nitric(V) acid manufacture, before the discharge of unusable gases to the atmosphere.

$$NO + NO_2 + 2NaOH \longrightarrow 2NaNO_2 + H_2O$$

In acidic conditions, the relative redox potentials are altered and nitrite ions disproportionate.

$$2NO_2^- + 2H^+ \longrightarrow NO + NO_2 + H_2O$$

Another example of disproportion is found in the reaction of nitrogen dioxide with water. (See p. 71.)

$$2NO_2 + H_2O \longrightarrow HNO_2 + HNO_3$$

6.2.8 **Odd-electron molecules**

The octet rule, of general use in predicting the structures of compounds, is not infallible. No matter how the electrons of nitrogen oxide and dioxide are grouped, one electron must remain unpaired, since they contain fifteen and twenty-three electrons respectively. Odd-electron molecules such as these often react in a way that will remove the unpaired electron, either by losing it (oxidation) or by pairing it with an unpaired electron in another molecule. In the case of nitrogen dioxide a dimer is formed (fig. 6.4).

When a bond is formed, energy is released. Applying Le Chatelier's principle dinitrogen tetroxide should therefore be more stable at low temperatures. This is confirmed by experiment; the liquid boiling at 10 °C is a dilute solution of NO_2 in N_2O_4, while the gas is a dilute solution of N_2O_4 in NO_2. Dissociation into NO_2 is not complete until 150 °C. Nitrogen oxide, NO, associates also, but only in the liquid state.

Odd-electron molecules are often coloured and paramagnetic. Nitrogen dioxide is brown, and the intensity of this colour deepens on heating the gas from 10 °C to 100 °C in a sealed flask. The monoxide is unusually colourless.

Chemically, the odd electron in these two oxides may be lost by oxidation (forming NO^+, nitrosonium, or NO_2^+, nitronium ions) or by reduction (forming NO^-, nitrosyl, or NO_2^-, nitrate(III), ions.

6.3 THE INDUSTRIAL AND AGRICULTURAL CHEMISTRY OF NITROGEN

On the experimental farm at Rothamsted, north of London, there is a plot of land which has grown a wheat crop every year since about 1840. Each year all the crop, including straw, has been removed. The plot is never fertilised. Nevertheless it continues to yield about 1000 kg ha^{-1} ($\frac{1}{2}$ ton/acre) each year. This contrasts with current British yields, from good land, of about 5000 kg ha^{-1}.

The difference lies in the fertiliser applied, since to produce the present-day crop about 100 kg ha^{-1} of nitrogen are necessary (together with 20 kg of phosphorus and 30 kg of potassium, and traces of other elements). The unfertilised plot relies on slow soil weathering and rainfall to supply its needs. Today's trend is to apply the required elements in the most concentrated form which can be absorbed by plants. Common fertilisers include ammonium nitrate(V) (34% N), diammonium hydrogen phosphate(V) (23% N), urea (45% N), and even pure liquid ammonia (82% N), the last being injected deeply into the soil. The ultimate in concentrated fertiliser is a dressing of red phosphorus, which has in fact been tried.

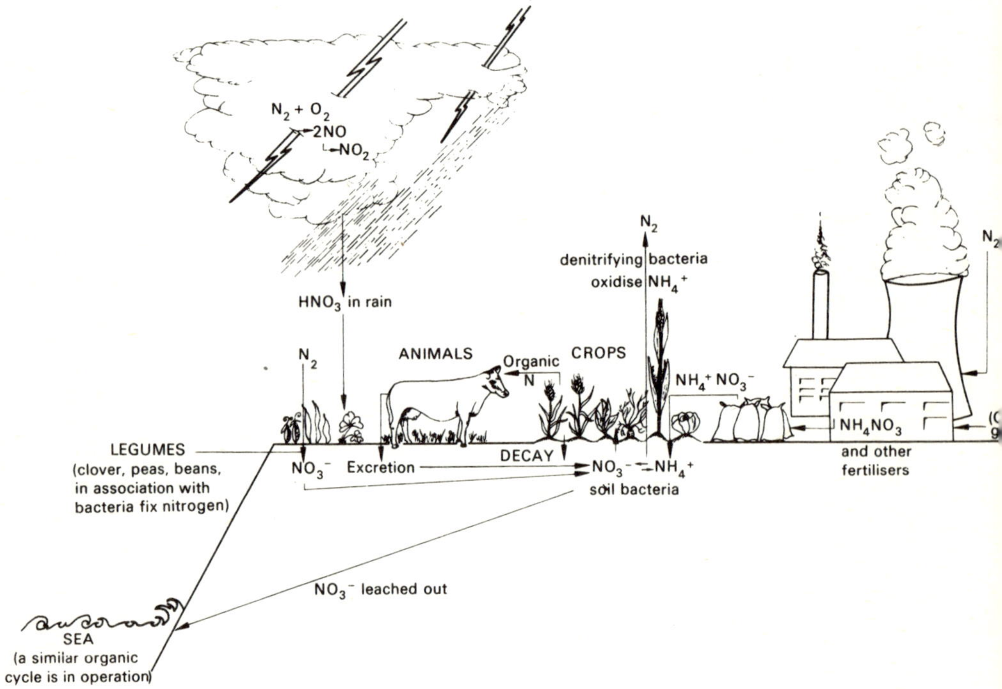

Fig. 6.8 Nitrogen cycle.

Thus, man has added to the natural nitrogen cycle, in which the nitrogen compounds needed for growth arise either from the decay of various plants or animals, or from the nitrogen-fixing ability of various bacteria living either free or in association with the roots of plants of the pea family. In the latter case, the nitrogen is reduced, probably with the aid of a molybdenum-containing enzyme.

Currently, the only economic method of converting atmospheric nitrogen to a compound is its reduction to ammonia. The hydrogen for this process is currently obtained from natural gas in Britain. In some countries, oil is used in a similar process. Coke and steam were formerly used, but the difficulties of handling coke are such that, even if coke were free, it would still be cheaper to use gas.

$$\left.\begin{array}{l}\underset{\text{natural gas}}{CH_4} + H_2O \longrightarrow CO + 3H_2 \\ \underset{\text{coke}}{C} + H_2O \longrightarrow CO + H_2\end{array}\right\} CO + H_2O \longrightarrow CO_2 + H_2$$

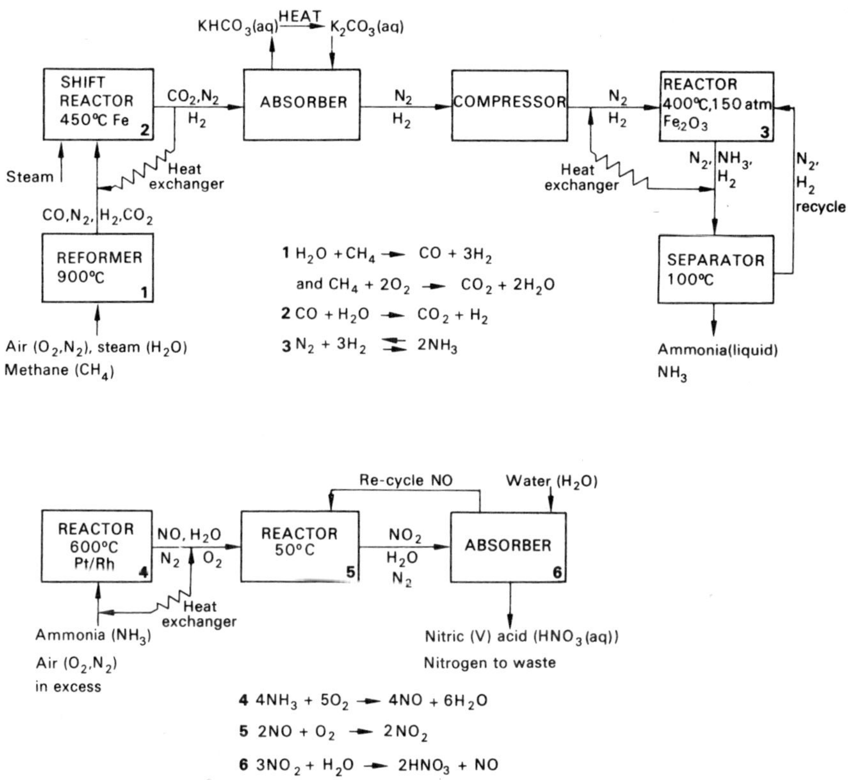

Fig. 6.9 The Haber process and the production of nitric(V) acid.

6.4 GENERAL POINTS OF COMPARISON FOR PHOSPHORUS, ARSENIC, ANTIMONY AND BISMUTH

The trends among this group of elements resemble those of group IV. You are referred to the last chapter for a discussion of the reasons behind the following trends.

6.4.1 Stability of the inert pair

Most binary compounds of the group V elements show decreasing stability of the +V oxidation state from phosphorus to bismuth. Some examples are tabulated below.

Halides

P	PBr_5 and PCl_5 known but dissociate to trihalide on heating
As	no pentahalides
Sb	$SbCl_5$ dissociates on heating
Bi	no pentahalides

Oxides

P_4O_{10}	stable to heat
As_2O_5	decomposes to As_2O_3 on heating
Sb_2O_5	decomposes to Sb_2O_3 on heating
Bi_2O_5	not known

The redox potentials for the reaction (V) state $\longrightarrow$ (III) state bear this out also.

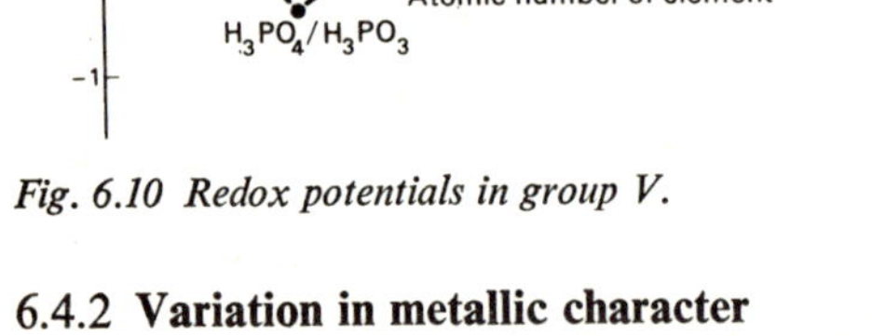

Fig. 6.10 Redox potentials in group V.

6.4.2 Variation in metallic character

Phosphorus, arsenic and antimony can be condensed from the vapour as covalently bonded X_4 molecules. P_4 is stable up to 250 °C, but Sb_4 is barely stable at room temperature. All these elements, and bismuth, also form a layer structure which becomes more symmetrical and metallic the lower the element is down the group.

	As acid (*typical of non-metallic oxides*)	*As base* (*typical of metallic oxides*)
P_2O_5	$+ H_2O$ (violent) $\longrightarrow H_3PO_4$ (phosphoric(V) acid)	–
P_2O_3	$+ H_2O$ (violent) $\longrightarrow H_3PO_3$ (phosphoric(III) acid)	–
As_2O_3	$+ H_2O$ (slow) $\longrightarrow HA_3sO_3$ (arsenic(III) acid)	+ conc. HCl $\longrightarrow AsCl_3$
Sb_2O_3	+ NaOH (dil.) $\longrightarrow Na_3SbO_3$ (sodium antimonate(III))	+ conc. HCl $\longrightarrow SbO^+$
Bi_2O_3	–	+ dil. HCl $\longrightarrow Bi^{3+}$(aq)

Table 6.3 Acidic and basic character of group V oxides.

The acidity of the oxides is summarised in table 6.3, which also illustrates the decrease in metallic character. The halides are hydrolysed in water to varying degrees. Phosphorus and arsenic trihalides are hydrolysed completely in excess water to form phosphoric(III) acid, and a hydrated arsenic(III) oxide respectively. This behaviour is characteristic of non-metal halides.

$$PCl_3 + 3H_2O \longrightarrow 3HCl + H_3PO_3$$
$$2AsCl_3 + 3H_2O \longrightarrow 6HCl + As_2O_3$$

Antimony and bismuth halides are only partially hydrolysed, and reversibly so. This behaviour is more 'metallic'.

$$Sb^{3+} + H_2O \rightleftharpoons SbO^+ + 2H^+$$

Solutions of antimony (and bismuth) salts have to be made up in dilute acid, in order to avoid the formation of SbO^+ and its precipitation as the oxychloride, SbOCl.

The hydrides decrease in thermal stability from phosphine to bismuthine. This instability is exploited in the classical test for arsenic, Marsh's test. The suspected sample is reduced to arsine, AsH_3, using zinc and hydrochloric acid. On passing the gas through a hot tube, a metallic mirror of arsenic is formed. If antimony is also present, a second mirror is formed in advance of the hot zone, as the more unstable stibine, SbH_3, decomposes first.

6.5 SUMMARY OF REACTIONS OF NITROGEN COMPOUNDS (refer to fig. 6.11)

Redox reactions

(a) By heating or in sunlight (a bottle on the windowsill goes brown).

$$4HNO_3 \longrightarrow 2H_2O + 4NO_2 + O_2$$

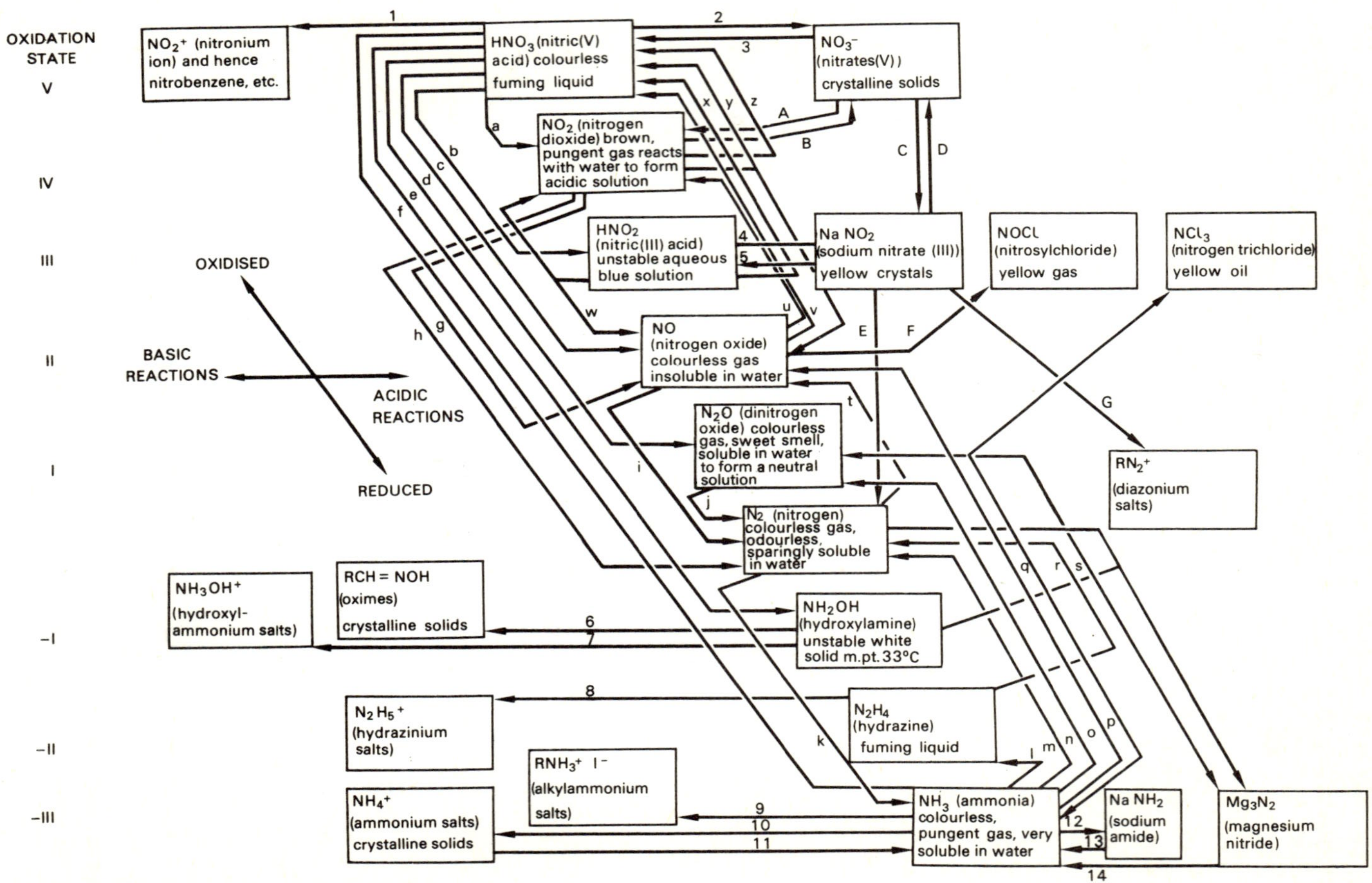

Fig. 6.11 Summary of the reactions of nitrogen.

By reducing agents forming covalent compounds: use hot, concentrated acid on carbon or sulphur.

$$4HNO_3 + C \longrightarrow 2H_2O + 4NO_2 + CO_2$$
$$6HNO_3 + S \longrightarrow 2H_2O + 6NO_2 + H_2SO_4$$

By reducing agents forming ions: use cold concentrated acid.

$$HNO_3 + H^+ + e^- \longrightarrow H_2O + NO_2$$
$$\text{Cu (copper turnings, preparative)} \longrightarrow Cu^{2+} + 2e^-$$
$$Fe^{2+}\text{ (iron(II) sulphate solution)} \longrightarrow Fe^{3+} + e^-$$
$$2I^-\text{ (potassium iodide solution)} \longrightarrow I_2 + 2e^-$$
$$H_2S\text{ (hydrogen sulphide solution)} \longrightarrow 2H^+ + S + 2e^-$$

(b) By heating sodium nitrate.
(c) By 50 per cent solution in water on copper turnings:

$$Cu \longrightarrow Cu^{2+} + 2e^-\text{ (preparative)}$$
$$HNO_3 + 3H^+ + 3e^- \longrightarrow 2H_2O + NO$$

(d) By heating ammonium nitrate: explosive in solid phase, controllable if diluted with water at all times.

$$NH_4^+ + NO_3^- \longrightarrow N_2O + 2H_2O$$

(e) By electrolysis of 50 per cent sulphuric acid, adding concentrated nitric acid slowly to the region around the mercury cathode.

$$HNO_3 + 6H^+ + 6e^- \longrightarrow NH_2OH + 2H_2O$$

(f) By boiling with excess dilute sodium hydroxide and aluminium foil.

$$NO_3^- + 6H_2O + 8e^- \longrightarrow NH_3 + 9OH^-$$
$$Al + 6OH^- \longrightarrow Al(OH)_6^{3-} + 3e^-$$

(g) Equilibrium at high temperatures. (See text.)
(h) Pass over hot copper to form copper(II) oxide (preparative).
(i) Pass over hot copper to form copper(II) oxide (preparative).
(j) Most substances which are flammable continue to burn. Hot copper can again be used.
(k) The Haber process, an industrial method only. (See text.)
(l) By cold sodium chlorate(I), using a gelatin catalyst.

$$2NH_3 + NaOCl \longrightarrow N_2H_4 + NaCl + H_2O$$

(m) By burning in oxygen, or passing over hot copper(II) oxide.

$$2NH_3 + 3CuO \longrightarrow N_2 + 3H_2O + 3Cu$$

(n) By heating with a nitrate. (See (d).)
(o) By burning in air over a platinum catalyst: industrial. (See text.)

$$4NH_3 + 5O_2 \longrightarrow 4NO + 6H_2O$$

(p) By treating ammonia with excess chlorine—a dangerous reaction. Better men than you have lost fingers this way! The product is hydrolysed by water.

$$NCl_3 + 3H_2O \longrightarrow NH_3 + 3HOCl\text{ (chloric(I) acid)}$$

(q) By adding most oxidising agents in solution (MnO_4^-, Fe^{3+}, I_2, Ag^+, etc.).

$$N_2H_4 \longrightarrow N_2 + 4H^+ + 4e^-$$

(r) By adding most oxidising agents in solution (MnO_4^-, Fe^{3+}, Ag^+).

$$2NH_2OH \longrightarrow N_2O + H_2O + 4H^+ + 4e^-$$

(s) By passing over very hot magnesium (or lithium, for lithium nitride).
(t) By passing air through an electric arc.
(u) Occurs spontaneously in air.
(v) By passing through most strong oxidising agents in acidic solution (MnO_4^-, $Cr_2O_7^{2-}$),

$$NO + 2H_2O \longrightarrow HNO_3 + 3H^+ + 3e^-$$

(w) Occurs spontaneously.

$$2HNO_2 \longrightarrow H_2O + NO + NO_2$$

Nitric(III) acid can be reduced to nitrogen oxide by iodide in acid solution.

$$e^- + H^+ + HNO_2 \longrightarrow H_2O + NO$$

(x) By adding most strong oxidising agents in acidic solution (MnO_4^-, $Cr_2O_7^{2-}$).

$$HNO_2 + H_2O \longrightarrow HNO_3 + 2H^+ + 2e^-$$

(y) Occurs spontaneously on solution in water.

$$3NO_2 + H_2O \longrightarrow 2HNO_3 + NO$$

(z) By passing through most strong oxidising agents in acidic solution (MnO_4^-, $Cr_2O_7^{2-}$).

$$NO_2 + H_2O \longrightarrow HNO_3 + H^+ + e^-$$

(A) By heating heavy metal nitrates(V). Lead nitrate(V) is used preparatively, being the only common anhydrous nitrate.

$$2Pb(NO_3)_2 \longrightarrow 2PbO + 4NO_2 + O_2$$

(B) By passing through oxidising agents in alkaline solution. (Compare with (z).)
(C) By heating sodium nitrate(V), either alone or with lead to form lead(II) oxide.
(D) By adding oxidising agents in alkaline solution. (Compare with (x).)
(E) By heating with an ammonium salt (e.g. chloride). The reaction is violent unless diluted with water.

$$NH_4^+ + NO_2^- \longrightarrow N_2 + 2H_2O$$

(F) By heating with chlorine: also analogues with fluorine, bromine. Salts containing NO^+ can be made by adding nitrosyl chloride to electron acceptors:

$$NOCl + SbCl_5 \longrightarrow NO^+SbCl_6^-$$

(G) By adding amines in acid solution. Consult an organic chemistry textbook for information regarding its significance in the preparation of dyes.

Acid–base reactions

(1) By mixing concentrated nitric acid with a very strong acid, for example, concentrated sulphuric(VI) acid, or chloric(VII) acid

$$2H^+ + HNO_3 \longrightarrow H_3O^+ + NO_2^+$$

(2) By the action of dilute nitric(V) acid on any common base (oxide, hydroxide, carbonate).
(3) By heating a nitrate(V) with a strong concentrated acid (e.g. concentrated sulphuric acid).

$$H^+ + NO_3^- \longrightarrow HNO_3$$

(4) By the action of dilute nitric(III) acid on any common base.
(5) By the action of any dilute strong acid on a nitrate(III) (e.g. dilute hydrochloric acid).
(6) By a reaction with aldehydes or ketones. For the importance of this reaction consult an organic chemistry textbook.
(7) and (8) By the action of any strong dilute acid. Reversible in base.
(9) By heating with a haloalkane in alcohol.

$$RI + NH_3 \longrightarrow RNH_3^+ I^-$$

(10) By the action of most dilute acids. Preparatively, the ammonia can be added in excess, and then evaporation removes this excess.

$$NH_3 + H^+ \longrightarrow NH_4^+$$

(11) By heating ammonium salts with strong bases. Preparatively calcium(II) hydroxide is used.

$$NH_4^+ + OH^- \longrightarrow NH_3 + H_2O$$

(12) By dissolving sodium in liquid ammonia.

$$Na + NH_3 \longrightarrow NaNH_2 + \tfrac{1}{2}H_2$$

(13) and (14) By the action of any weak acid, e.g. water.

$$NH_2^- + H_2O \longrightarrow NH_3 + OH^-$$
$$N^{3-} + 3H_2O \longrightarrow NH_3 + 3OH^-$$

6.6 USES OF PHOSPHORUS, ARSENIC, ANTIMONY AND BISMUTH

Although the use of phosphorus as its sulphide, P_4S_3, and as the red allotrope in matches is well known, this accounts for very little of the phosphorus produced industrially by the following electrically heated process:

$$\underset{\text{phosphate rock}}{2Ca_3(PO_4)_2} + \underset{\text{coke}}{10C} + \underset{\text{quartz rock}}{6SiO_2} \longrightarrow \underset{\text{calcium silicate}}{6CaSiO_3} + 10CO + \underset{\text{yellow phosphorus}}{P_4}$$

Most of the phosphorus is burnt in air to form phosphorus(V) oxide, which in turn is reacted with water to form very pure phosphoric(V) acid. This pure form is used for treating iron against rust in car manufacturing, etc., and for incorporation as the sodium salt in baking powder. A larger quantity of phosphoric acid still is produced by treating calcium phosphate rock with concentrated sulphuric acid and is then converted to ammonium phosphate, triple superphosphate, calcium hydrogenphosphate and other fertilisers, or to water softeners for incorporation in washing powders.

The uses of the remainder of the group depend mainly on their toxic or metallic properties. Arsenates are common constituents of insecticides and timber preservatives (most motorway fencing will give a strong positive test for arsenic). Arsenic and antimony find their way into bearing alloys, and antimony and bismuth into low-melting alloys.

QUESTIONS

1 Using examples taken only from group V, show how the presence of the following types of bonding features can be used to predict the properties of the compound which contain them:
(a) a strong covalent bond,
(b) a weak covalent bond,
(c) a non-bonding pair,
(d) ions,
(e) a co-ordinate bond,
(f) a hydrogen bond.

2 Explain, in terms understandable to the non-scientist,
(a) why his prosperity depends on the discovery that iron oxide catalyses the reaction between nitrogen and hydrogen;
(b) why aerial dusting of pastures with red phosphorus is beneficial, and not particularly dangerous;
(c) why the naturally occurring isotope of bismuth ($^{209}_{81}Bi$) can be sold freely as an ingredient of stomach powders even though radioactive;
(d) how Marsh's test for arsenic, as used in the best Agatha Christie novels, works;
(e) why so many explosives have names beginning with 'nitro-'.

3 Explain in technical terms
(a) why nitrogen dioxide is more strongly coloured than its dimer, dinitrogen tetroxide;
(b) why the Science Research Council is unlikely to grant funds for attempts to prepare nitrogen pentachloride;
(c) why medieval instructions for the forging of a sword are reputed to include, as the final stage, the plunging of the sword, red-hot, through a living slave;
(d) why the oxides of phosphorus are given the formula P_4O_n, n=6, 8, 10, etc.;
(e) why ammonia, unlike methane or phosphine, is readily soluble in water;

(f) why 'phosphorus(V) oxide' dissolved in water produces a more acidic solution than antimony(V) oxide;
(g) why there is a greater proportion of nitrogen dioxide present at high temperatures in the equilibrium $2NO_2 \rightleftharpoons N_2O_4$;
(h) why red phosphorus, although a giant lattice, vaporises at about 250 °C;
(i) why the reaction of NH_4Cl with KNH_2 is called an acid–base reaction;
(j) why N^{3-} is substantially larger than Bi^{3+}.

4 Describe, with a flow diagram, the production of ammonia from nitrogen and hydrogen by the Haber process. Explain
(a) why the temperature chosen is higher than that which gives a maximum yield;
(b) why the pressure chosen is lower than that which gives a maximum yield;
(c) why the hydrogen is obtained from oil and steam rather than from coal and steam;
(d) to what uses the product is put, both directly and via nitric acid.

5 Calculate
(a) the value of the platinum catalyst used in the oxidation of ammonia (assume gauze 1 mm thick, conversion rate 10^{-5} mole NH_3 cm^{-2} sec^{-1}, designed input 17×10^4 kg NH_3 day^{-1}, cost of Pt 2×10^3 £kg^{-1}, density of Pt 20 gcm^{-3});
(b) the volume of methane required to supply the above plant (assuming 100 per cent conversion in the cracking reaction);
(c) the mass of nitric acid produced (assuming 90 per cent overall yield based on ammonia);
(d) the rise in temperature of the gas stream at the oxidation stage ($4NH_3 + 5O_2$ (as air) $\longrightarrow 4NO + 6H_2O$, $\Delta H = -903\,kJmol^{-1}$, heat capacities in $Jmol^{-1}K^{-1}$: N_2, 31.5; H_2O, 39.1; NO, 32.6).

6 Write three balanced equations for each of the following instances:
(a) the reduction of a nitrogen compound other than an oxide,
(b) a disproportion,
(c) the hydrolysis of a non-metal chloride,
(d) a compound of phosphorus acting as an acid,
(e) a precipitation reaction involving a compound of nitrogen.

7 The following statements are not correct in all respects. There may be exceptions, ambiguities, omissions or deliberate mistakes. Write out a correct version of the statement and comment on the theoretical background to it.
(a) The evolution of ammonia from a solution treated with aluminium foil and sodium hydroxide proves the presence of a nitrate(V).
(b) The +V oxidation state is a better oxidising agent as one goes from nitrogen to bismuth.
(c) Non-metal chlorides are hydrolysed to form hydrogen chloride.
(d) A co-ordinate bond is the result of a proton accepting a lone pair.
(e) Endothermic compounds are unstable.
(f) Disproportion occurs under acidic conditions.
(g) Artificial fertilisers are required for healthy crops.

7

Oxygen and Sulphur

A case of first-element dissimilarity

7.1 INTRODUCTION

In several groups there are marked differences between the first, or head, element and the rest of the group. When oxygen is compared with sulphur (or with the less important elements selenium, tellurium and polonium) these differences are quite sharp and there are relatively few points of similarity. The differences arise from causes already discussed:

(1) the relatively small size of the head atom (oxygen) or its ions

Element	*Radius of atom*/nm	X^{2-} *ion*
Oxygen	0.075	0.140
Sulphur	0.105	0.185

(2) the fact that the outer electrons of sulphur are in the third shell ($3s^23p^4$), which can be expanded to accommodate a total of eighteen electrons.

7.2 OXIDE AND SULPHIDE IONS

7.2.1 **Lattice effects**

Both oxygen and sulphur have six electrons in their valence shell (oxygen $1s^2$, $2s^22p^4$; sulphur $1s^2$, $2s^22p^6$, $3s^23p^4$). As expected, they acquire an extra two electrons in many reactions to make up the octet. Although this is an endothermic process, many ionic oxides and sulphides exist, the necessary energy for the ionisation being recovered by the very high values of lattice energy conferred by the small size and double charge. Sulphides, although requiring less energy for the corresponding process ($S(s) + 2e^- \longrightarrow S^{2-}(g)$), recover comparatively little of this when a lattice is formed, since the sulphide ions are large, and cannot pack closely to cations. There are many cases of sulphides being more cova-

Ionisation of the element entails:	*Oxygen*	*Sulphur*
Atomisation of the standard state $\frac{1}{2}O_2(g) \longrightarrow O(g)$ or $S(s) \longrightarrow S(g)$	+248	+223
Combination with the first electron $X(g) + e^-(g) \longrightarrow X(g)$ *exothermic*	−142	−200
Combination with the second electron $X^-(g) + e^-(g) \longrightarrow X^{2-}(g)$ *endothermic*	+844	+532
Total input of energy	+950	+555
Outweighed by lattice energy CaX (cf. $CaCl_2$, 2230)	−3500	−2950

Table 7.1 Formation of oxide and sulphide ions. (*All values in* $kJmol^{-1}$.)

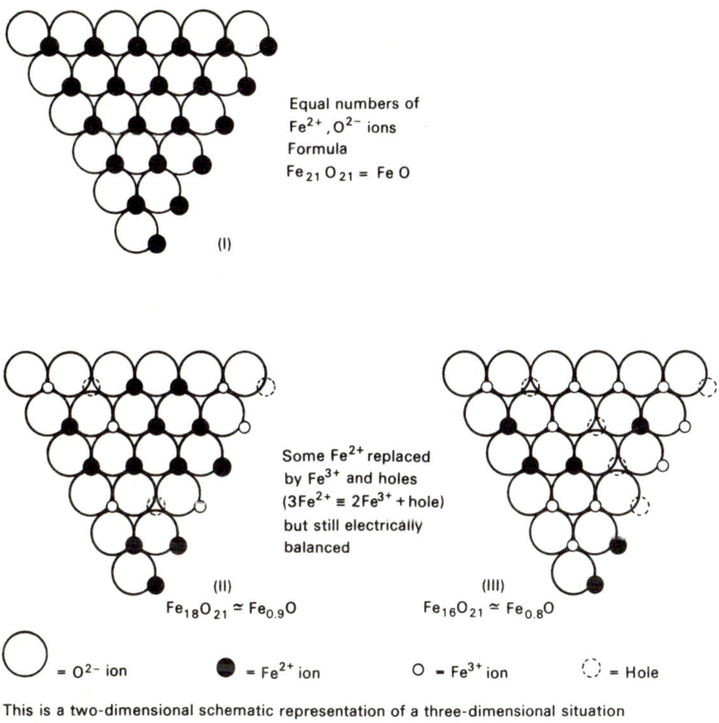

Fig. 7.1 Non-stoichiometric structures.

lent in character than the matching oxides. For example, cobalt and nickel oxides dissolve readily in acid to form the appropriate salts:

$$O^{2-} + 2H^+ \longrightarrow H_2O \quad (CoO \longrightarrow CoCl_2)$$

but the sulphides are covalent and do not react with dilute acids.

Although oxides and sulphides are small compared to other anions, they are much larger than the average cation. The crystal is often composed of a lattice of close-packed spherical oxide ions (imagine a pyramid of billiard balls) with the gaps filled by the cations. In such cases, not all gaps need be filled, as fig. 7.1 illustrates. The oxides of metals with two valencies are often non-stoichiometric, that is, they possess no definite formula. The holes in the lattice are very good absorbers of all frequencies of radiations, so that such compounds can be recognised by their black or brown appearance. The same considerations apply to sulphides, although the covalent bonding requires directional bonds which cannot fit the 60° angles needed for close-packing.

7.2.2 **Geological occurrence**

Oxygen is very abundant in the earth's atmosphere, so that many of the ores in the earth's crust are oxides and oxy-anions. The presence of oxygen in the atmosphere is principally due to the photosynthetic activity of plants which, over millions of years have converted a predominantly reducing atmosphere into an oxidising one.

$$xCO_2 + yH_2O \xrightarrow[\text{synthesis}]{\text{photo-}} Cx(H_2O)y + xO_2$$

There is a loose relationship between the position of an element in the periodic table and the nature of its most common ore.

7.2.3 **Acid–base chemistry**

The acid–base properties of oxides are inextricably connected with the presence of oxygen in water, our dominant solvent. Thus, oxides are usually classified according to their behaviour in the aqueous solvent system.

(i) They may react with acid, oxide ions being weak bases.

$$O^{2-} + 2H^+ \longrightarrow 2H_2O$$

This reaction applies to the more ionic oxides, those of metals with low charge and high ionic radius. Such oxides are said to be basic, and include Na_2O, CaO, FeO and CuO.

(ii) They may react with bases.

$$CO_2 + OH^- \longrightarrow HCO_3^- \ (\xrightarrow{OH^-} CO_3^{2-})$$

This reaction applies to the more covalent oxides, those of non-metals which can form covalent bonds to oxygen. Such oxides are said to be acidic and include SO_2, P_2O_5, NO_2 and I_2O_7.

(iii) They may do both, in which case they are described as amphoteric. Metals in high oxidation states, and metals to the right-hand side of the periodic

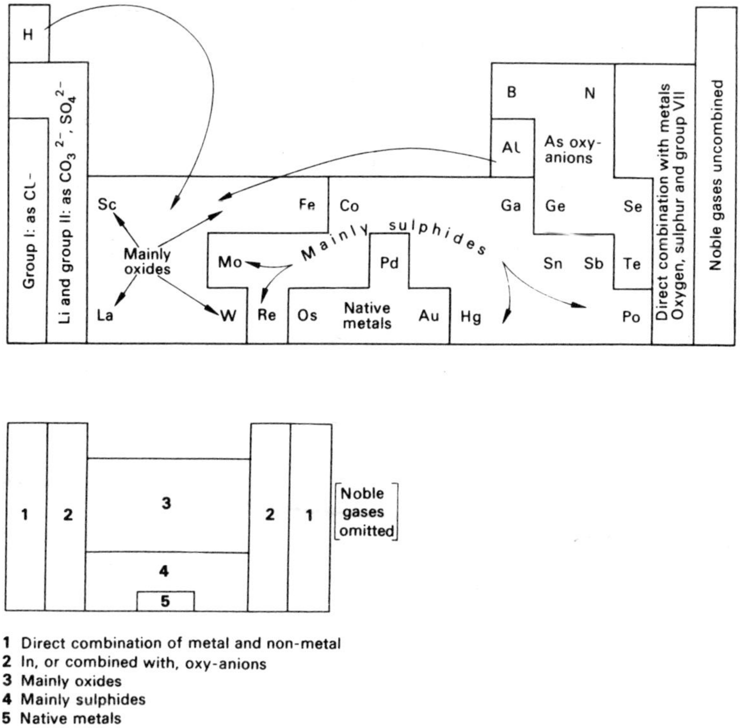

Fig. 7.2 Stable ores and the periodic table. This table is a simplification: elements lying near boundaries are often found in the alternative form (e.g. Fe *is mainly found as* Fe_2O_3, *haematite, but also as pyrites,* FeS_2*). Notice the loose symmetry involved.*

table, generally form oxides in this class, the high charge 'formally' associated with the metal ion attracting additional oxygen ligands.

$$Al_2O_3 + 6HCl \xrightarrow[\text{heat}]{\text{conc.}} 2AlCl_3 + 3H_2O \ (\longrightarrow 2Al(H_2O)_6^{3+})$$

$$Al_2O_3 + 6OH^- + 3H_2O \xrightarrow[\text{heat}]{\text{conc.}} 2Al(OH)_6^{3-}$$

Others in this group include ZnO, Cr_2O_3, PbO and SnO_2.

This classification is not a rigid one, and oxides form a continuous range from basic (dissolving only in acids) to those with a little amphoteric character (dissolving in alkali with great difficulty) to fully amphoteric (dissolving equally readily in acid or alkali) and so through to basic.

7.2.4 **Oxidising properties**

Oxides are also classified by their redox behaviour. 'Normal' oxides are those formed when the element burns in air, and have no particular oxidising or reducing properties. Sub-oxides contain a smaller proportion of oxygen, and

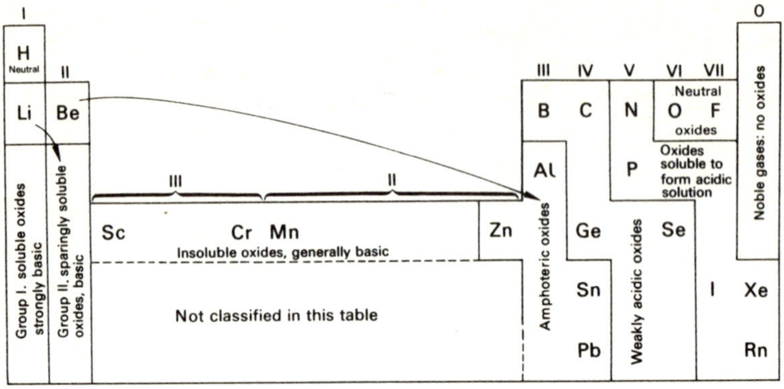

The oxides included have the oxidation state at the top of each column.

Fig. 7.3 Acid/base classification of oxides and the periodic table.

have to be prepared by special methods. For example, heating metal oxalates generates a reducing atmosphere of carbon monoxide and the oxide. Pyrophoric (catching fire in air) iron(II) oxide may be prepared in this way.

$$FeC_2O_4 \longrightarrow FeO + CO + CO_2$$

Higher oxides contain a metal in a higher oxidation state than normal, and are usually good oxidising agents. For instance, lead(IV) oxide can be prepared by the action of sodium chlorate(I) on lead(II) salts, and will oxidise hydrogen chloride to chlorine, or set fire to hydrogen sulphide.

$$Pb^{2+} + OCl^- + H_2O \longrightarrow PbO_2 + Cl^- + 2H^+$$
$$PbO_2 + 4HCl \longrightarrow PbCl_2 + 2H_2O$$

Other examples include MnO_2, CrO_3 and V_2O_5.

Other types of oxides arise in the following way. The oxygen molecule can acquire electrons in stages, thus

$$\underset{\text{gaseous molecule}}{O_2} \longrightarrow \underset{\text{superoxide ion (burn K in air} \rightarrow KO_2)}{O_2^-} \longrightarrow \underset{\text{peroxide ion (burn Na, Ba in air)}}{O_2^{2-}(\bar{O}\text{–}\bar{O})} \longrightarrow \underset{\text{oxide ion}}{O^{2-}}$$

Peroxides and superoxides are good oxidising agents. Peroxides react with acids, unlike higher oxides, to give hydrogen peroxide which decomposes to oxygen with the heat of reaction.

$$O_2^{2-} + 2H^+ \longrightarrow H_2O_2 \longrightarrow H_2O + \tfrac{1}{2}O_2$$

Sulphides are not classifiable in the same way. They react only with acid ($S^{2-} + 2H^+ \longrightarrow H_2S$), not with alkali. The sulphide ion is a good reducing agent ($S^{2-} \longrightarrow S + 2e^-$) so that it can never be found with a metal in a high oxidation state (cf. Fe^{3+}, chapter 11). Disulphide ions do exist (see the next section).

7.3 DOUBLE AND SINGLE BONDS

The bond energies (BE) for the single and double covalent bonds of oxygen and sulphur are given in table 7.2.

Element	*Single bond*	*Double bond*	*Enthalpy* X = X ⟶ (X–X)
Oxygen	150	500	+200
Sulphur	260	360	−160

Table 7.2 Bond energies in group VI. (All values in kJmol^{-1}.)

The reason for the extra stability of the π-bond of a first-row element is explained under carbon (section 5.4). In this case the differences affect the state in which the element occurs. The enthalpy of reaction of X=X molecules to (–X–X–) units in a polymer can be calculated from the bond energies involved (table 7.2). Conversion in this direction involves breaking the X=X double bond, and forming an X–X single bond together with a half-share of the single bonds to the neighbouring units. Hence

enthalpy of reaction = BE (X=X) − [BE (X–X) + 2 × $\frac{1}{2}$ BE(X–X)]. The results are given in table 7.2. They show clearly that oxygen should exist as the double-bonded diatomic molecule which, being non-polar with a molecular mass of 32, must be a gas; and that sulphur should be a single-bonded polymer, which, because of a high molecular mass, should be a solid.

In fact the situation for sulphur is more complex. Various rings and chains form different allotropes at different temperatures, as shown in table 7.3. Even so, S_2 (S=S) becomes stable at very high temperatures, in accordance with Le Chatelier's principle. The comparative stability of the single bonds also show themselves in the formation of poly-anions. For oxygen only peroxide ions are known, and these are good oxidising agents ($E^\ominus$ in acid, + 0.7 V), capable of converting sulphide into sulphur, iodide to iodine, or chromium(III) hydroxide to chromate(VI). Sulphur, on the other hand, forms a series of salts:

$$^-S{-}S^-,\ ^-S{-}S{-}S^-,\ ^-S{-}S{-}S{-}S^-,\ \text{and so to } S_6^{2-}.$$

All of these can be isolated from a solution of sulphur in sodium sulphide, and are very weakly oxidising.

The extra strength of the double bond in oxygen, is significant in the 'un-reactivity' of oxygen. Although all elements except halogens and noble gases react directly with oxygen, these reactions are very slow at room temperature. An initial source of high temperature such as a match flame will complete enough reaction for the very large enthalpy of reaction to set off the rest. This large enthalpy is a result of the small atom of oxygen being able to form short, and therefore strong, bonds. Sulphur, on the other hand, occurs as weakly bonded atoms and also forms weakly bonded molecules. Therefore it reacts more readily but less violently than oxygen. At room temperature it reacts with mercury to form mercury(II) sulphide—a good way of dealing with mercury spills. There is

Temperature °C	0°	96°	119° 200°	444°
Appearance	Rhombic (diamond) crystals	Monoclinic (needle) crystals	amber ← black liquid → brown	Vapour
Molecular form	S_8 ring	S_8 (differently stacked to rhombic form)	mixture of S_8 rings; S_∞ long chains; S_n short chains	Mixture of S_8, S_6, S_4, S_2
Comments	Crystallises at low temp, e.g. from solution in CS_2	Crystallises at high temp., e.g. from melt	Maximum viscosity at approx. maximum chain length. Can be 'frozen' by pouring into water, when rubbery solid, plastic sulphur is formed. It is unstable and decays to rhombic crystals	Molecular size decreases with temperature. If 'frozen' by contact with cold surface, an amorphous mixture, 'flowers of sulphur' is formed
		Change can be observed by expansion in volume in S/oil mixture	True m.pt. If heated rapidly without time to change to monoclinic melts at 113 °C	

Table 7.3 Allotropes of sulphur.

also an account of a French geologist of the eighteenth century who buried a few hundred kilograms of an iron and sulphur mixture in his garden to test his theory of the formation of volcanoes. He was not disappointed! In general, though, heat is required to initiate the reaction.

7.4 d ORBITALS AND SULPHUR CHEMISTRY

Oxygen, having the electronic structure 2 : 6, must form bonds using electrons from the second principal shell only. The limited number of protons in its nucleus are unable to hold any electrons in a third shell, and the second shell is limited to eight electrons. Hence a covalency of two is the maximum possible, to which may be added a co-ordinate bond.

Sulphur forms bonds in the third principal shell. This can hold a maximum of eighteen electrons, and compounds of sulphur are known with up to twelve electrons in the outer shell.

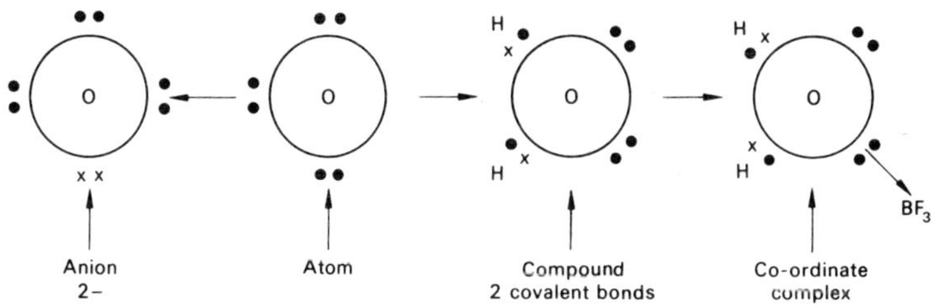

Fig. 7.4 Bonds to oxygen atoms.

	anion, 2⁻; 8 electrons in outer shell
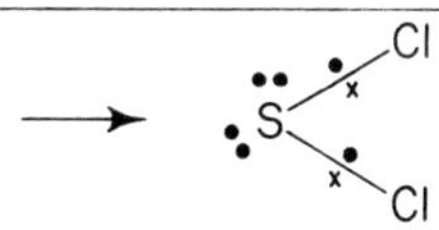	2 covalent bonds providing 8 electrons in outer shell– "stable octet" also, e.g. H_2S
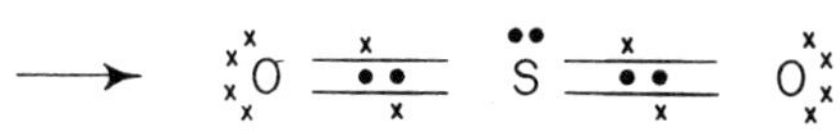	10 electrons in outer shell also, e.g. SO_3^{2-}, SCl_4
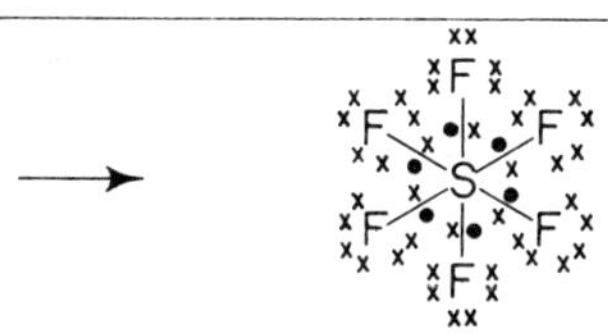	12 electrons in outer shell also, e.g. H_2SO_4, SO_2Cl_2, SO_3

Table 7.4 Bonds to sulphur atoms.

The extra electrons are accommodated in the d orbitals. One other difference in bonding is that oxygen can form hydrogen bonds. These account for the unusual properties of water, and are discussed fully in chapter 10.

7.5 INDUSTRIAL CHEMISTRY OF SULPHUR: SULPHURIC ACID

Although several compounds of sulphur are useful industrially, the industrial chemistry of sulphur is overshadowed by the production of sulphuric acid. It is the most common industrial acid, and nearly 4000 million tonnes are used in the United Kingdom annually (about 60 tonnes per head of population!).

The first stage in its manufacture is the production of sulphur dioxide. In any plant designed solely for sulphuric acid production, the source of sulphur is always the element. For instance, British firms import molten sulphur from the Texas deposits, where it occurs in underground beds overlain by quicksands. To save the expense of sinking a shaft in poor ground, a triple borehole is sunk. Air and water at 180 °C under pressure are pumped down two of the concentric pipes. The molten sulphur rises up the third of the pipes as a froth with air and

Formula	*Compound*	*Uses*
S	Sulphur	Vulcanising rubber, fumigant, insecticide (on burning)
SO_2	Sulphur dioxide	Preservative (peas), petrol and sugar refining, bleach
$Ca(HSO_3)_2$	Calcium hydrogensulphite	Production of paper pulp
H_2NSO_3H	Sulphamic acid	Acid cleaner (e.g. brewing and food processing equipment)

Table 7.5 Uses of sulphur compounds

water. The liquid sulphur is burnt to sulphur dioxide, much as if it were oil in a furnace.

$$S + O_2 \longrightarrow SO_2$$

Other sources of sulphur must be allied to the production of other, more valuable, chemicals to be economic. Zinc is extracted from zinc blende (ZnS) by roasting the blende in air.

$$ZnS + \tfrac{3}{2}O_2 \longrightarrow ZnO + SO_2$$

and the zinc oxide is then used to make zinc. Pyrites (FeS_2) and lead or copper ores are treated similarly. Hydrogen sulphide occurs in some sources of natural gas, and, if its concentration is more than a few parts per million, it is extracted in an organic solvent. The cost of this operation is recouped by boiling the solvent and burning the hydrogen sulphide in a limited supply of air.

$$H_2S + \tfrac{1}{2}O_2 \longrightarrow H_2O + S$$

This represents the principal source of sulphur in France.

Sulphur dioxide emitted from power stations and factory chimneys has been considered as a source of sulphur. It arises from sulphur impurities in fuels, particularly coal. However, the problem of treating the large volume of gas has not yet been solved economically, so that the cost would have to be borne by an anti-pollution subsidy. Since sulphur dioxide is a severe irritant to animals and deadly to vegetation, a subsidy might be justifiable.

The sulphur dioxide is then oxidised to sulphur trioxide. The direct reaction is slow and a catalyst must be used. Formerly this was done in the 'lead chamber' process, but this process is now uneconomic, principally because of the impure acid it produced. Oxides of nitrogen were used as oxygen carriers:

$$\left.\begin{array}{l} NO_2 + SO_2 \longrightarrow NO + SO_3 \\ NO + \tfrac{1}{2}O_2 \longrightarrow NO_2 \end{array}\right\}\text{with several side reactions}$$

In the contact process now in use, a mixture of sulphur dioxide and air is passed over a vanadium(V) oxide catalyst.

$$SO_2 + \tfrac{1}{2}O_2 \rightleftharpoons SO_3 \quad (\Delta H = -95\ \text{kJmol}^{-1}).$$

This is an equilibrium reaction, favoured by high pressure and low temperature. In practice, the expense of installing compressors outweighs the profits of the extra yield, and the temperature is kept at about 500 °C in order to maintain a reasonable rate of reaction. About 98 per cent conversion is nevertheless achieved. This process was patented in the early nineteenth century, but the catalyst then in use, platinum, combined irreversibly (was 'poisoned') with impurities, such as arsenic(III) oxide from sulphide ores. Vanadium(V) oxide is resistant to poisoning, but the gas stream still has to be free from dust, adding to the expense.

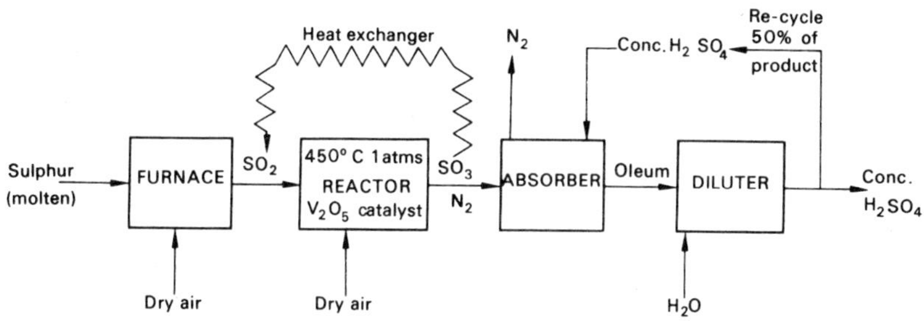

Fig. 7.5 Manufacture of sulphuric acid.

The sulphur trioxide is not dissolved in water, since it forms an intractable acid mist. Instead, it is first dissolved in concentrated sulphuric acid to form 'oleum', which is then diluted with water. Some of the acid formed is then re-cycled (fig. 7.5).

7.6 PROPERTIES OF SULPHURIC ACID

The properties of the concentrated acid are useful in a number of different ways:

(a) It combines vigorously with water to form a strong acid, completely dissociated in dilute solution.

$$H_2SO_4 + 2H_2O \longrightarrow 2H_3O^+ + SO_4^{2-}$$

For the typical properties of acids see chapter 10. Industrial uses of dilute sulphuric acid include the cleaning of rust from steel to be welded, the production of ammonium sulphate fertiliser and the extraction of titanium(IV) oxide for paints.

$$3H_2SO_4 + Fe_2O_3 \longrightarrow Fe_2(SO_4)_3 + 6H_2O$$
$$2\,NH_3 + H_2SO_4 \longrightarrow (NH_4)_2SO_4$$

(b) In the concentrated form, protons may be donated to bases other than water. In the laboratory, hydrogen chloride is often prepared by warming sodium chloride with the acid: in industry phosphoric acid for fertilisers is prepared in the same way.

$$NaCl + H_2SO_4 \longrightarrow NaHSO_4 + HCl$$

$$Ca_3(PO_4)_2 + 3H_2SO_4 \longrightarrow 3CaSO_4 + 2H_3PO_4$$

(c) The concentrated acid also acts as a dehydrating agent. This type of reaction occurs in two stages: first, a protonation (cf. examples (a) and (b) above), followed by the elimination reaction. Alcohols form alkenes, and sugars are dehydrated to carbon.

$$\underset{\text{ethanol}}{CH_3CH_2OH} \longrightarrow \overset{H}{\overset{|}{C}}H_2{-}CH_2{-}\overset{\oplus}{O}H_2 \longrightarrow \underset{\text{ethene}}{CH_2 = CH_2} + H_3O^+$$

$$C_m(H_2O)_n \text{ (general formula for sugar)} \longrightarrow {}_mC + {}_nH_2O$$

The formation of the nitronium (NO_2^+) ion which is involved in the nitration of benzene and its derivatives involves a similar process.

$$HNO_3 \longrightarrow H_2NO_3^+ \longrightarrow H_2O + NO_2^+ \longrightarrow H_3O^+ + NO_2^+$$

The nitronium ion attacks toluene to form the explosive trinitrotoluene. Similar reactions are important in the formation of many yellow and red azodyes.

(d) The concentrated acid is an oxidising agent. When it reacts with metals, it is reduced to sulphur dioxide, in contrast to the dilute acid which produces hydrogen. This type of reaction has no industrial applications.

$$H_2SO_4 + 2H^+ + 2e^- \longrightarrow SO_2 + 2H_2O$$

$$M \longrightarrow M^{2+} + 2e^-$$

Concentrated sulphuric acid also oxidises halide ions (section 12.5). It reacts directly with benzene derivatives to form sulphonic acids.

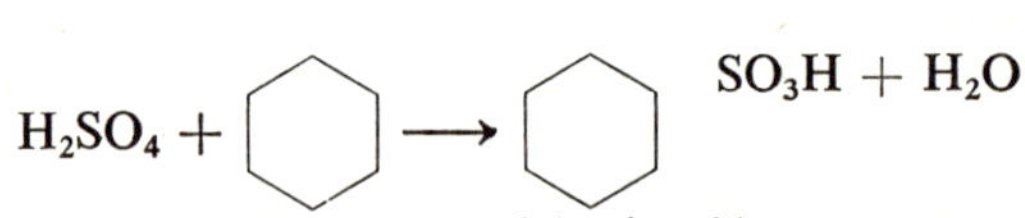

benzene sulphonic acid

Sodium salts of alkylbenzene sulphonic acids (e.g. $C_{12}H_{23}-$ 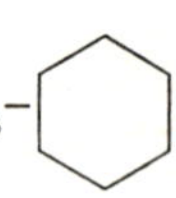 SO_3Na) are used as detergents (washing-up liquids).

7.7 SUMMARY OF SULPHUR CHEMISTRY

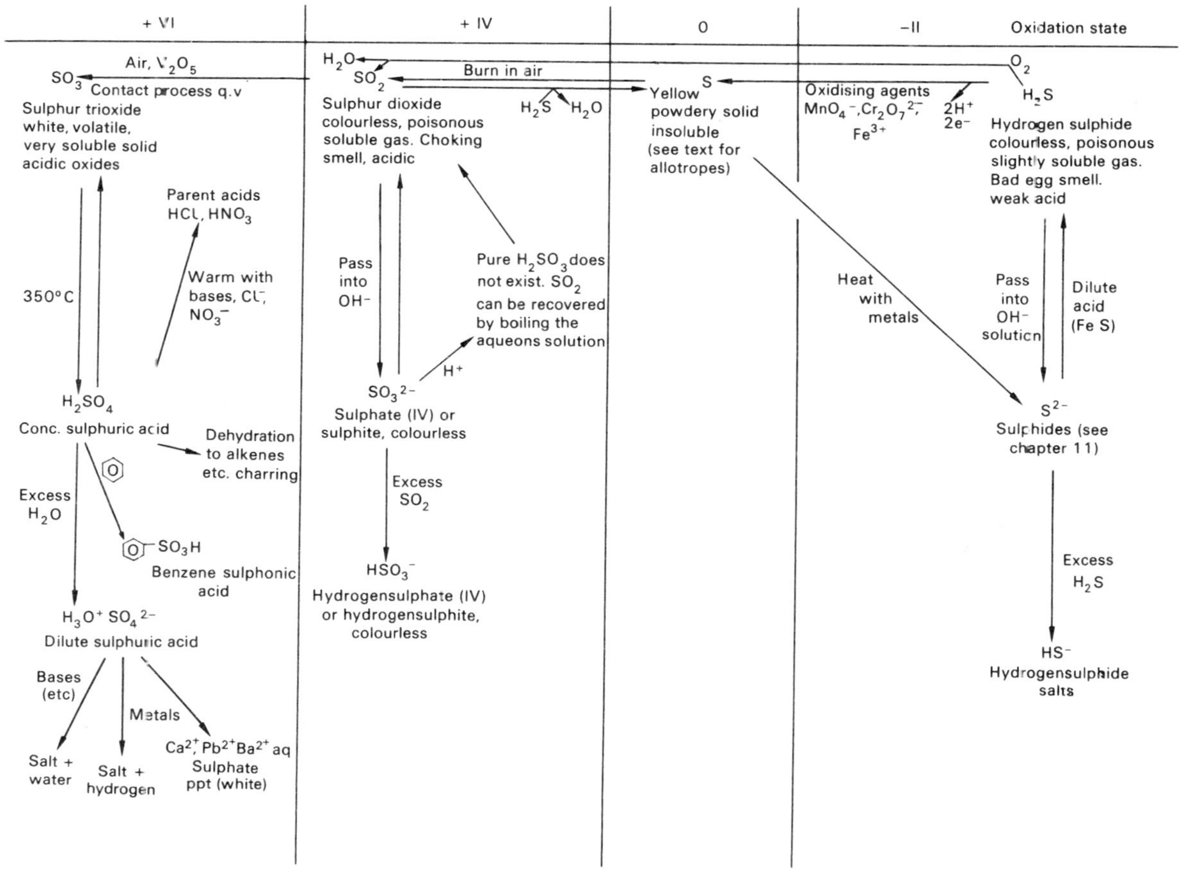

7.8 SUMMARY OF OXYGEN CHEMISTRY

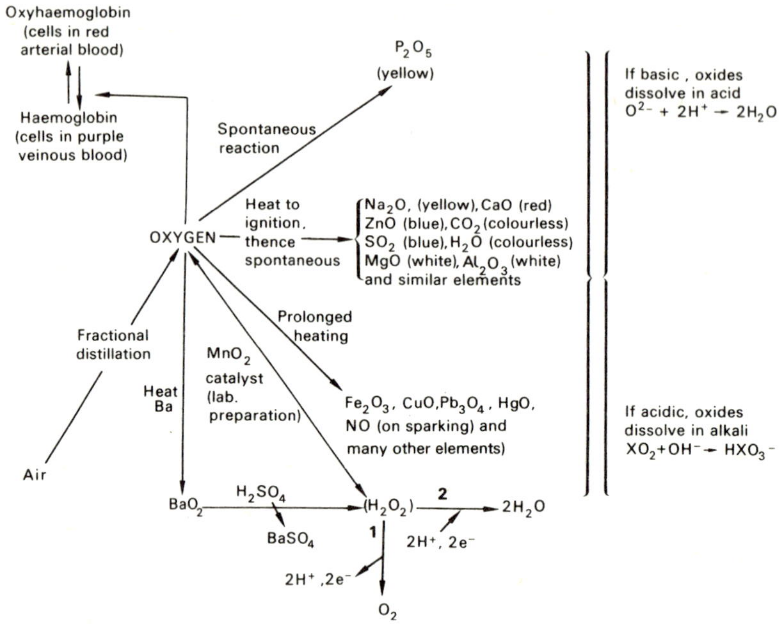

QUESTIONS

1 Show how the comparatively small size, and the limit of eight electrons in the outer shell, of the oxygen atom cause the following differences between oxygen and sulphur:
(a) the greater ionic character of most oxides,
(b) the relative boiling points of water and hydrogen sulphide,
(c) the existence of fluorides SF_n, where $n > 2$,
(d) the physical states of the elements,
(e) any one other difference of your choice.

2 Trace the chemical reactions which are used industrially to convert sulphur into sulphuric(VI) acid by the Contact Process. Account for the following:
(a) the choice of equilibrium conditions at the $SO_2 \longrightarrow SO_3$ stage;
(b) zinc sulphide—which has to be mined and paid for—competes with sulphur as a raw material in this process but sulphur dioxide—which is emitted free from power station chimneys—does not compete;
(c) the process was patented in 1798 specifying platinum as catalyst but was not commercially successful until a hundred years later.

3 Write balanced equations for two examples of each of the following cases:
(a) sulphuric(VI) acid acting as an oxidising agent,
(b) hydrolysis,
(c) an oxide dissolving in sodium hydroxide,
(d) oxidation by a dioxide,
(e) a sulphide acting as a reducing agent.

4 Explain the following:
(a) Transition metal oxides are often black, or strongly coloured.
(b) The process $X(g) \longrightarrow X^{2-}(g)$ is more endothermic for oxygen than for sulphur: but the heat of formation of an oxide is usually larger than that of the corresponding sulphide.
(c) Non-metal oxides are usually acidic.
(d) Sulphur dissolves readily in solutions of sulphides.
(e) Concentrated nitric acid can act as a base.
(f) Sulphur dioxide is a bent molecule.

5 Explain, as if to a person with a smattering of science:
(a) how a lorry driver came to die when he tipped his acid waste on to a dump of sulphide waste;
(b) why extensive felling of the equatorial forest might (but is unlikely to) affect the breathability of our air;
(c) why an oil-burning power station does less damage to vegetation than a coal-burning one, and where this damage is most likely to occur;
(d) why the boiling point of water is so exceptional, and what importance this has for life in general;
(e) why magnesium hydroxide is good for some types of indigestion;
(f) what the fire brigade are likely to do if called to an accident involving a tanker load of concentrated sulphuric acid.

6 The following statements are not correct in all respects. There may be exceptions, ambiguities, omissions or deliberate mistakes. Write out a correct version of the statement and comment on the theoretical background to it.
(a) Metal oxides are basic.
(b) The number of atoms in the molecules of allotropes of sulphur increases with temperature.
(c) Oxygen forms a maximum of two bonds.
(d) Hydrogen halides may be prepared by the action of sulphuric acid on halide ions.
(e) The partial pressure of natural oxygen is $\frac{1}{5}$ atmosphere.

8
The Halogens

A group in which redox potentials and bond energies are dominant

8.1 INTRODUCTION

The halogens—fluorine, chlorine, bromine and iodine, with the artificial and highly radioactive element astatine—form a well-known and well-ordered group in the periodic table. All have seven electrons in their outer shell, leaving room for one low-energy electron (to complete the 'stable octet') before higher orbitals have to be filled. Consequently, the chemistry of each element resembles that of the others, and is very much concerned with the gaining of this electron, whether acquired outright to form ionic compounds or by sharing to form covalent compounds. Some of the more important properties of the halogens are given in table 8.1. (X in this chapter is used to denote any halogen atom.)

All the halogens, and many of their compounds, have physiological effects. Sodium chloride (common salt) is a basic item of diet, and traces of fluoride and iodide are also needed for strong teeth and for growth hormone respectively. However, more than traces of sodium or hydrogen fluorides can damage tissue by blocking repair mechanisms: hydrogen fluoride burns generally require skin grafts. All the elements are pungent, choking and toxic if inhaled, and, except for iodine, blister the skin on contact. Remember that chlorine has been used as a war gas, and treat it and the other halogens with respect.

8.2 PROPERTIES RELATED TO REDOX POTENTIALS

8.2.1 **General**

The standard electrode, or redox, potential is a convenient way of measuring the energy change when electrons are gained or lost in reations in an aqueous medium. In the case of the halogens, it refers to the process

$$\tfrac{1}{2}X_2 + e^- \longrightarrow X^-$$

The values for the various halogens are given in table 8.1, and it is obvious that fluorine has by far the largest potential, i.e. it is the best oxidising agent. The reason is shown in fig. 8.1.

	Fluorine	Chlorine	Bromine	Iodine
Relative atomic mass (common isotopes)	19 (19)	35.5 (35, 37)	80 (79, 81)	127 (many)
Electronic structure	2:7	2:8:7	2:8:18:7	2:8:18:18:7
Density at melting point/gcm^{-3}	1.0	1.5	3.1	4.9
Boiling point/K	85	238	332	457
Electron affinity/$kJmol^{-1}$	−348	−364	−342	−314
Electronegativity	4.0	3.0	2.8	2.5
Covalent radius/nm	0.07	0.10	0.115	0.13
Ionic radius X^-/nm	0.14	0.18	0.195	0.215
Bond energy X–X/$kJmol^{-1}$	158	242	193	151
Bond energy H–X/$kJmol^{-1}$	562	431	366	299
Hydration enthalpy X^-/$kJmol^{-1}$	506	364	335	293
Lattice energy NaX/$kJmol^{-1}$	902	771	733	684
Redox potential ($\frac{1}{2}X_2 \longrightarrow X^-$)/V	+2.87	+1.36	+1.07	+0.54

Table 8.1 The halogens.

Bond energy decreases in the order chlorine, bromine, iodine, owing to the increasing length of the X–X bond. At the same time, the electron affinity, which is the term opposite to it in the diagram, also decreases, because the

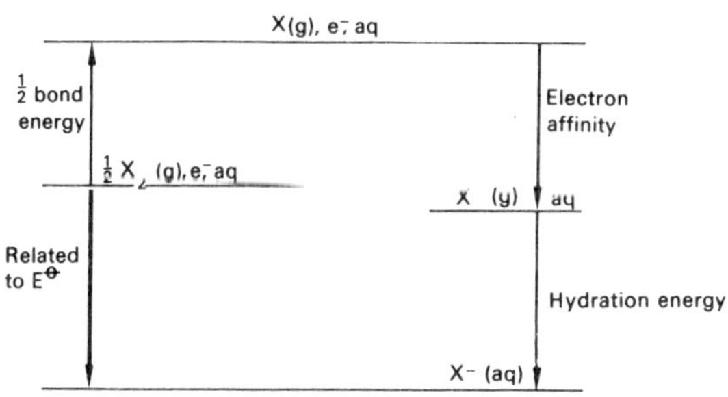

Fig. 8.1 Analysis of redox potentials for halogens.

Note: as an approximation bromine and iodine are assumed to be gaseous and, strictly, heats of atomisation should be used in place of bond energies for these.

electron gained is shielded from the attraction of the positive nucleus by the extra electrons present. In fact, the figures for the enthalpy of the combined process, $\frac{1}{2}X_2(g) + e^- \longrightarrow X^-(g)$, are very similar:

$$X{=}Cl,\ -243\ kJmol^{-1};\ Br,\ -246;\ I,\ -239$$

Fluorine is a special case. Contrary to the general trend, the bond in F_2 is much weaker than the bond in Cl_2. This is probably due to the repulsion of non-bonding pairs on each fluorine atom at each end of the short F–F bond. However, and again contrary to the trend, the electron affinity is also small for fluorine. This is again probably due to the non-bonding electrons repelling the incoming electron, which has to approach them closely on account of the small size of fluorine. When these two oppose each other, the net result is

$$\tfrac{1}{2}F_2(g) + e^- \longrightarrow F^-(g),\ \Delta H = -269\ kJmol^{-1}$$

This value is still comparable with the other halogens.

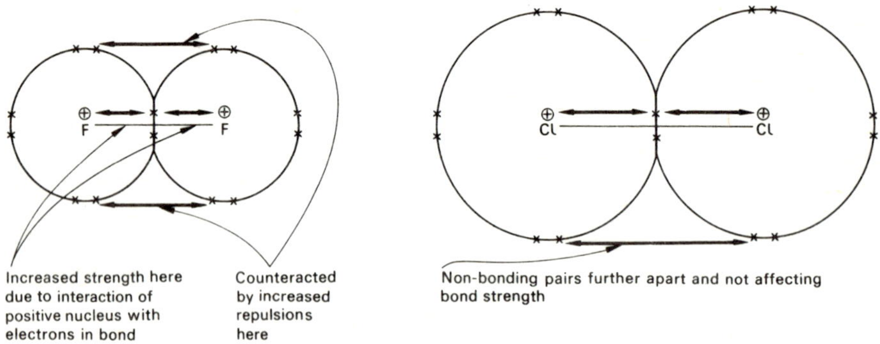

Fig. 8.2 Bond strength of halogens.

The difference in redox potentials is therefore due almost entirely to the difference in hydration energies. Here fluorine wins hands down owing to the small size of the fluoride ion (see table 8.1). The increased charge density allows very strong interaction with the positive end of the dipole of the water molecule.

8.2.2 **The laboratory preparation of the elements**

Halogens can be prepared from halides by oxidation. This involves choosing a reagent which has a suitable redox potential, and which will not contaminate the product.

In the case of fluorine, which has $E^\ominus = +2.87\,V$, there is no known oxidant of higher potential, so that the preparation from fluoride must be carried out electrolytically, as described in the industrial section, below. Chlorine, however, possesses $E^\ominus = +1.36\,V$, so that chlorides or hydrochloric acid can be used as starting material. A convenient method is to drop the concentrated acid on to potassium manganate(VII) crystals ($E^\ominus = +1.52\,V$), whereupon chlorine is evolved in the cold. The more traditional method of preparation is to heat the concentrated acid with manganese dioxide ($E^\ominus = +1.23\,V$). At first sight this potential is not enough to oxidise chlorides: in fact, the two values are sufficiently close for an equilibrium to be set up. From this the chlorine is progressively expelled by heating, until the reaction goes to completion. The gas should be passed through brine to remove the (more soluble) hydrogen chloride impurity and collected by downward delivery.

$$4HX + MnO_2 \longrightarrow MnX_2 + X_2 + 2H_2O \ (X = Cl, Br, I)$$

Manganese dioxide is also suitable for oxidising bromides ($E^\ominus = +1.07\,V$), although it is more convenient to generate the hydrogen bromide by dropping concentrated sulphuric acid on to a mixture of sodium bromide and manganese dioxide. The bromine can be collected as a liquid using a Liebig condenser. Iodide solutions ($E^\ominus = +0.54\,V$) can conveniently be oxidised by concentrated nitric acid ($E^\ominus = +0.81\,V$), from which the solid iodine can be separated by filtering and washing.

8.2.3 **Manufacture of elements**

The industrial preparation of the halogens is subject to the same considerations as their laboratory preparation, except that a greater emphasis is placed on the cost of materials. Chlorides are not oxidised by expensive manganese compounds, but by electricity or air instead. Bromides are oxidised using chlorine. Iodides do not occur in large deposits in nature; sodium iodate(V) occurs mixed with

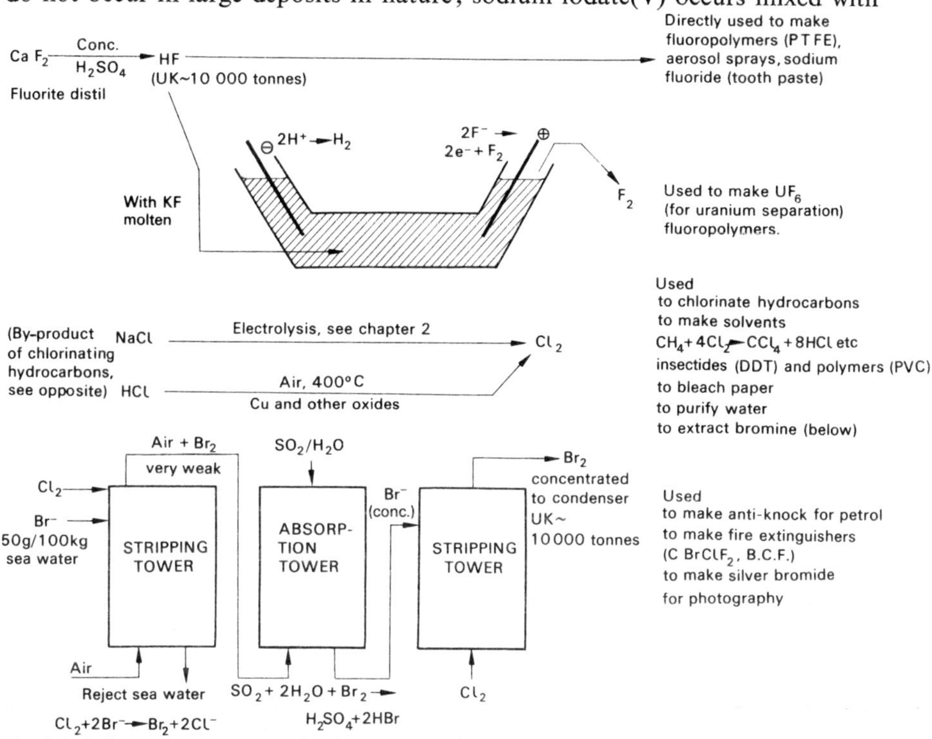

Fig. 8.3 Industrial preparation of halogens.

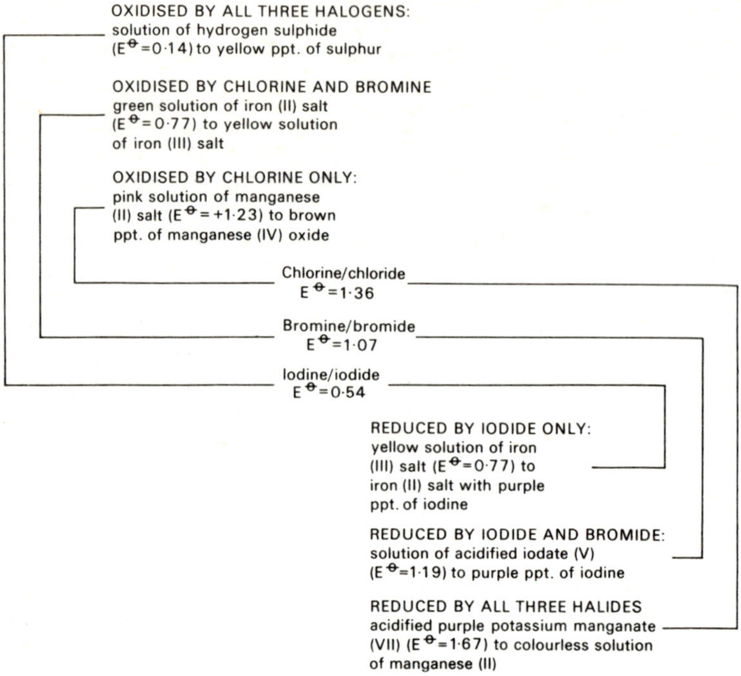

Fig. 8.4 Redox reactions of halogens and halides.

sodium nitrate in Chile. These ores are separated by fractional crystallisation and the iodate(V) is reduced with sulphate(IV).

8.2.4 **Preparation of hydrogen halides**

Given a supply of the salts, the hydrogen halides can be liberated by treating these with a non-volatile and stronger acid.

$$X^- + H^+ \longrightarrow HX$$

Concentrated sulphuric acid can be used with fluorides and chlorides, but is too strong an oxidising agent to be used with bromides and iodides, since it converts them to bromine and iodine. In these cases concentrated phosphoric acid can be used. Since all the hydrogen halides are gases, very soluble in water, they can be collected by downward displacement.

Hydrogen halides can also be prepared by the action of water on non-metal halides. Phosphorus trihalides are commonly used. The trichloride is stable and readily available, but the bromide and iodide are prepared *in situ*. For instance, to prepare hydrogen bromide, bromine is dropped on to moist red phosphorus. Phosphorus tribromide is formed but is hydrolysed immediately:

$$P + \tfrac{3}{2}Br_2 \longrightarrow \underset{\text{phosphorus tribromide}}{PBr_3} \qquad PBr_3 + 3H_2O \longrightarrow \underset{\text{phosphoric (III) acid}}{H_3PO_3} + 3HBr$$

$$\text{Compare} \quad PCl_3 + 3H_2O \longrightarrow H_3PO_3 + 3HCl$$

In either case the reaction is vigorous and needs to be cooled.

8.2.5 **Aqueous redox reactions**

These can be easily inferred from a table of redox potentials. It is possible to select substances which will be oxidised by the halogen, or reduced by the halide, in any of the combinations shown in fig. 8.4.

Fluorine, of course, oxidises all other halides; chlorine oxidises bromides and iodides; bromide oxidises iodides; and iodine will not displace any of the others. If chlorine in tetrachloromethane is shaken with a suspected halide solution, then if chloride is present there is no reaction and no change in the organic layer; if bromide is present, the released bromine turns the organic layer brown; if iodide is present, the organic layer turns purple. The purple colour in tetrachloromethane is the true colour of iodine; the brown colour seen in aqueous solutions is due to interaction between iodine and water.

8.3 OXY-ACIDS OF THE HALOGENS

Fluorine forms no oxy-acids, since its electronegativity is too high to allow the non-bonding electrons to form a co-ordinate bond with oxygen atoms. As the electronegativity decreases, so the thermal stability of the other oxy-acids and salts increases.

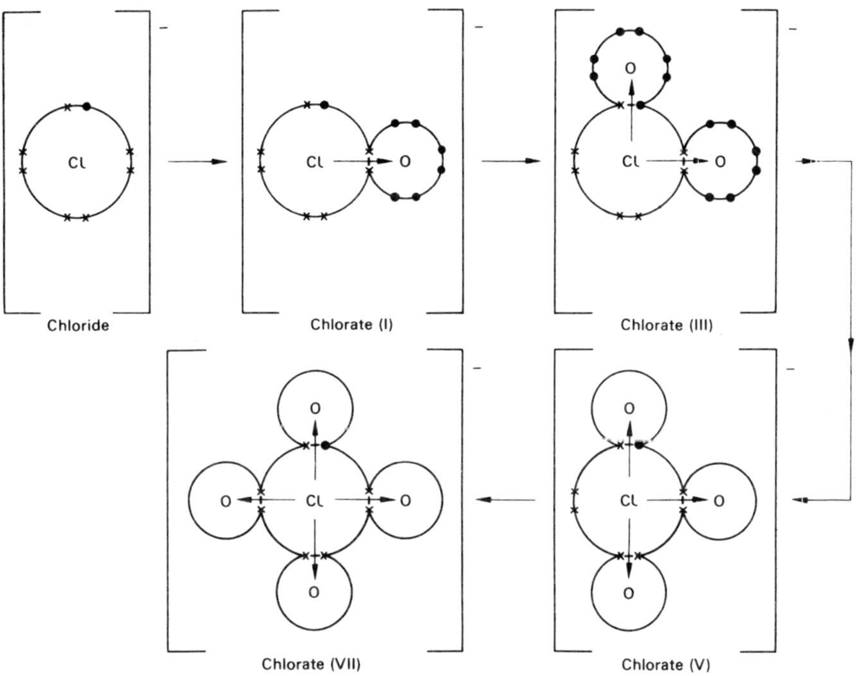

All the possible oxy-anions are known, with the exception of bromate(III)—although bromate(VII) has only recently been made by a 'cheat' synthesis from β-decay of radioactive selenate(VI). As might be expected, for compounds with such a high proportion of electronegative elements, they are all strong acids (except halate(I)) and strong oxidising agents ($E^{\ominus}$ always above $+1.20$ V in acid solution). The bleaching action of chlorine is due to the formation of chloric(I) acid in solution:

$$Cl_2 + H_2O \longrightarrow HCl + HOCl$$

and bleaching powder contains calcium chlorate(I).

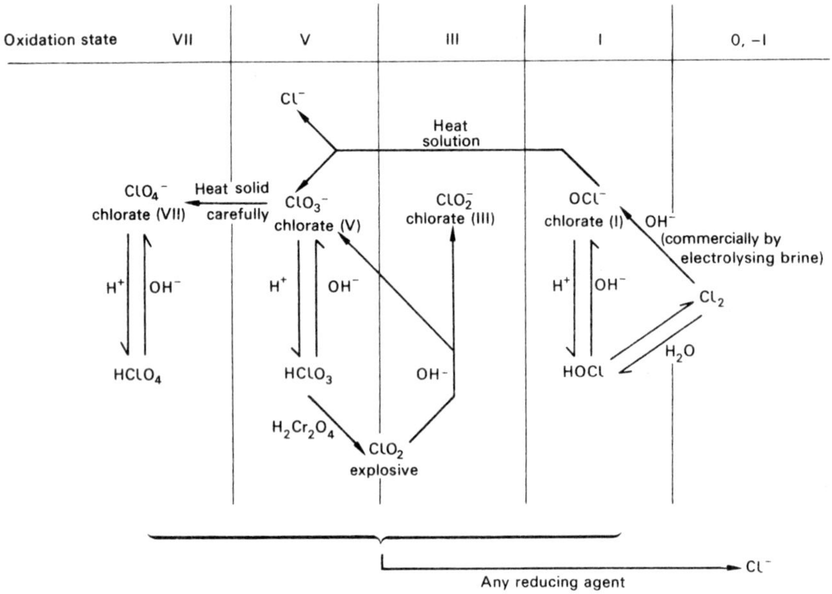

Fig. 8.5 Reactions of chlorates.

8.4 PROPERTIES RELATED TO THE LATTICE ENERGY OF HALIDES

The Born–Haber cycle describing the formation of halide salts is shown in fig. 8.6. According to the argument presented in section 8.2.1, stages 3 and 4 do not affect the comparative stability of the halides: and stages 1 and 2 are the same whichever halide of a given metal is being considered.

Therefore the relative stability of, say, the sodium halides depends *only* on the relative magnitude of the lattice energy. Because the fluoride ion is much smaller than any of the others, sodium fluoride has a much greater lattice enthalpy, and a larger enthalpy of formation. For this reason, fluorine will displace chlorine from a chloride crystal, and so on for the remaining elements.

If a salt with a metal in an unusually high oxidation state (e.g. silver(II)) is required, the same considerations suggest the fluoride. Only this will have sufficiently high lattice energy to compensate for the extra ionisation energy required to convert Ag^+ to Ag^{2+}. Silver(II) fluoride is known, as are copper(III) fluoride and other anomalous fluorides.

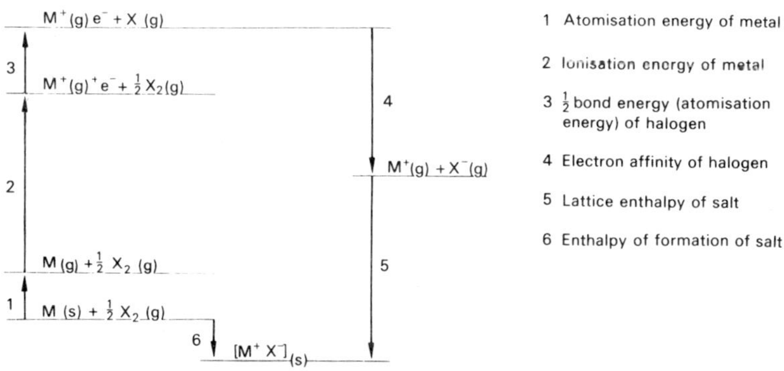

Fig. 8.6 Born–Haber cycle for halides.

Solubility depends on ionic size also. The solubilities of the group I and group II halides are discussed in section 3.7. However, the same argument applied to silver halides does not yield correct predictions, because, unlike group I, silver forms some covalent halides. Silver fluoride is ionic and soluble. The difference between the experimental lattice energy and the one calculated for an ionic lattice show the other silver halides to be increasingly covalent towards silver iodide—which, by reason of its covalency, is insoluble. For the application of solubilities to tests for halides, see section 12.6.

8.5. PROPERTIES RELATED TO BOND ENERGIES

8.5.1 **General**

When halogens react with another element or compound, two bond energies are important from a comparative point of view—the bond energy of the halogen and the bond energy of the new bond that is formed. Consider, for example, the reaction with hydrogen. The general principle is that bonds become weaker as they become longer. A glance at table 8.1 confirms this for the H–X bonds—quite strikingly for the H–F bond. As previously mentioned, however, (section 8.2) fluorine is an exception to the rule for X–X bonds.

For the reaction $H_2 + X_2 \longrightarrow 2HX$, the overall enthalpy is equal to

$$(\text{Bond energy } H_2 + \text{Bond energy } X_2) - 2(\text{Bond energy HX}).$$

The effect of the low bond energy of fluorine is twofold. Firstly, to break the bond requires little energy: fluorine atoms are easily formed: consequently the rate of reaction is much increased. Secondly, the total enthalpy of reaction—the heat released—is greater than the general trend would predict. The reactions of fluorine therefore tend to be rapid and violent. A mixture of hydrogen and fluorine explodes spontaneously. Chlorine, on the other hand, releases quite a lot of energy, but requires more energy than fluorine before the reaction will

	F	Cl	Br	I
Bond energy H_2	436	436	436	436
Bond energy X_2	158	242	193	151
2 × *Bond energy* HX	2 × 562	2 × 431	2 × 366	2 × 299
Enthalpy of reaction	−530	−184	−103	+11

Table 8.2 Enthalpy of reaction $H_2 + X_2 \rightarrow 2HX$.

start. A mixture of hydrogen and chlorine is stable until irradiated (with ultra-violet light) or ignited. Bromine occupies an intermediate position between chlorine and iodine: a mixture of hydrogen and bromine vapour will burn more or less quietly once ignited. By the time iodine is reached, the strength of the HX bond has decreased to such an extent that the reaction becomes reversible. Curiously, the lower activation energy for iodine enables its reaction to start spontaneously, but then the smaller amount of heat evolved prevents the reaction becoming violent.

Similar effects appear in reactions with other elements and compounds, as shown in table 8.3.

8.5.2 **Formation of high oxidation states in fluorine compounds**

Another trend shown by the above table is the tendency of fluorine to form compounds in high oxidation states. For covalent compounds, to convert (say) PX_3 to PX_5 involves expanding the original octet of electrons to ten. The extra

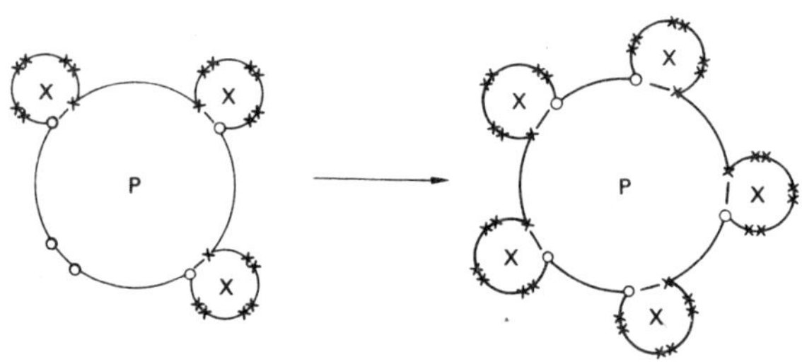

Fig. 8.7 Expansion of 'stable' octet.

electrons must enter the d orbitals of the shell at a higher energy, and therefore the extra bonds are rather weak. This weakness is further enhanced by the crowding of five halogen atoms around one phosphorus. Fluorine escapes the

Reactant	*Fluorine*	*Chlorine*	*Bromin*	*Iodine*
Carbon (charcoal)	Ignites spontaneously $\longrightarrow CF_4$	No reaction	No reaction	No reaction
Sulphur powder	Ignites spontaneously $\longrightarrow SF_6$	On heating forms S_2Cl_2	On heating forms S_2Br_2	No reaction
Red phosphorus	Ignites spontaneously $\longrightarrow PF_5$	On heating forms PCl_3 and PCl_5	Spontaneously forms PBr_3	Spontaneously forms PI_3
Iron filings (other metals similar)	Ignites if just warm $\longrightarrow FeF_3$	Reacts on heating $\longrightarrow FeCl_3$	Reacts if warmed $\longrightarrow FeBr_3$	Reacts if warmed $\longrightarrow FeI_2$
Ammonia	Explosive reaction $\longrightarrow N_2 + HF$	Burns $\longrightarrow N_2 + HCl$	Reacts $\longrightarrow N_2 + HBr$	Reacts $\longrightarrow NI_3$
Water	Forms ozone and F_2O (difluorine monoxide)	Reversible reaction to form HCl and HOCl	Reversible reaction to form HBr and HOBr	No reaction
Methane	Explosive reaction $\longrightarrow CF_4$, HF + some C	Explosive on irradiation $\longrightarrow CCl_4$, HCl, $CHCl_3$ etc. + some C	Slow reaction $\longrightarrow CH_3Br$, HBr	No reaction

Table 8.3 Reactions of halogens.

worst of both effects: the first by forming stronger bonds and the second by its smaller size.

8.5.3 **Acidity of HX**

All the hydrogen halides are acids, but hydrogen fluoride is the only weak acid. This is often ascribed to the bond energy of H–F being so large as to prevent dissociation, but further analysis shows this to be an over-simplification. The increase in bond energy in HF is almost cancelled out by the increase in hydration energy. If the enthalpy change for $HX(g) \longrightarrow H^+X^-(aq)$ is calculated, then the value for HF is -1144 kJmol^{-1}: for HCl, -1135 kJmol^{-1}. The probable reason for the weakness of hydrogen fluoride as an acid is that its true concentration is lowered by a competing equilibrium: besides

$$HF(aq) \rightleftharpoons H^+(aq) + F^-(aq)$$

more hydrogen fluoride is taken up by

$$F^-(aq) + HF(aq) \rightleftharpoons HF_2^-(aq) \quad (F---H---F).$$

This ion is held together by hydrogen bonds, which are fully discussed together with their effect on boiling points, in 10.2.2.

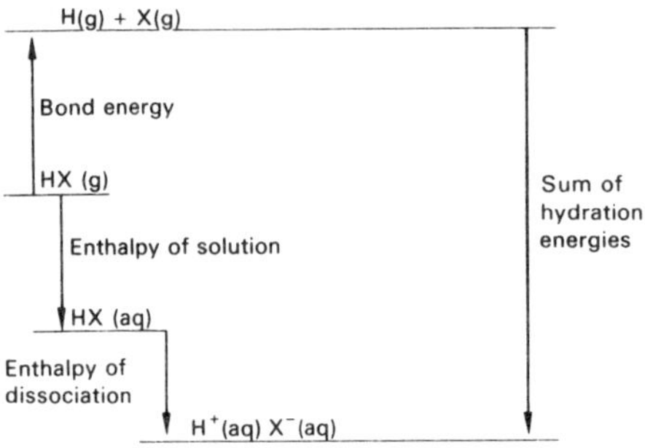

Fig. 8.8 Enthalpy cycle for acidic dissociation of hydrogen halides.

8.6 SUMMARY OF CHLORINE CHEMISTRY

Other halogens follow this general pattern, and for exceptions and comparisons you are referred to the chapter.

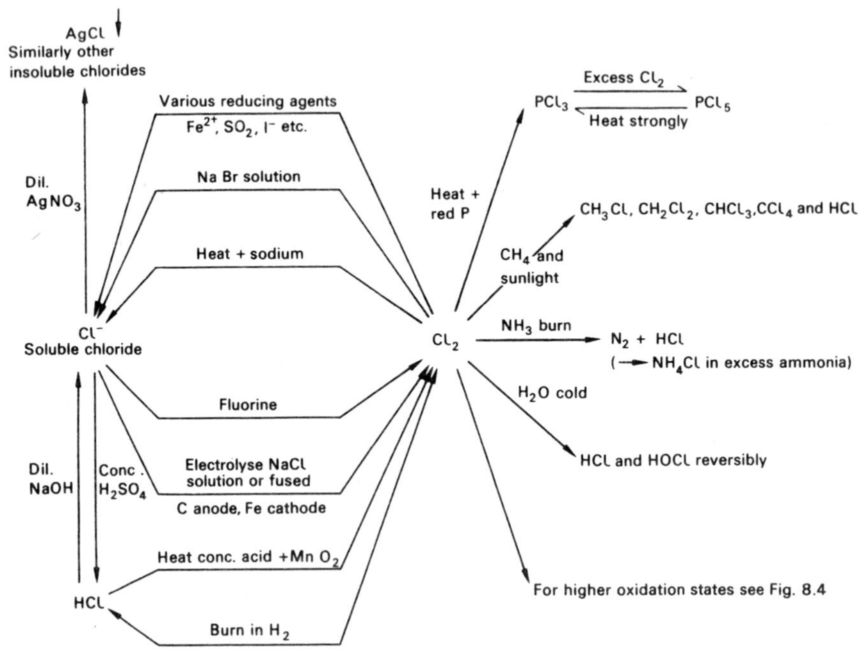

QUESTIONS

1 How do the following properties vary through the sequence F–Cl–Br–I? Account for the variation and in each case trace at least one instance of an observable property related to it. Show how these chosen properties are examples of the trends and similarities in the halogen family.
(a) electron affinity,
(b) electronegativity,
(c) ionic radius,
(d) bond enthalpy X–X,
(e) bond enthalpy H–X,
(f) redox potential X_2/X^-,
(g) any one other property of your choice.

2 Explain the following statements:
(a) Manganese dioxide, although having a smaller redox potential than chlorine, will nevertheless oxidise hydrogen chloride to chlorine.
(b) Elements in a high oxidation state are normally combined with fluorine, not with the other halides.
(c) Iodine reacts spontaneously with phosphorus and many metals, but the reaction with chlorine—though more exothermic—needs igniting.
(d) Chlorine, passed into excess potassium iodide, in aqueous solution, turns the solution brown: but if carbon tetrachloride is present, that layer becomes purple.
(e) There is no one general method for the preparation of hydrogen halides.

3 Assign values to the relative ecological advantages and disadvantages of the halogen products given below. Would you, as legislator, encourage or discourage their use? Or would you call for further research on their effects? If the former, state why; if the latter, outline the results you would want before a definite decision was made.
(a) 1,1-di(4-chlorophenyl)-2,2,2-trichlorethane(DDT)
For: good insecticide, malaria and other insect-borne diseases virtually wiped out.
Against: decline in viability of predator birds' eggs owing to accumulation in food chain; most strains of housefly now immune.
(b) Dichlorodifluoromethane.
For: useful aerosol propellant.
Against: known to react with ozone: there is an ozone layer around the earth which absorbs harmful ultra-violet radiation.
(c) Polychlorinated biphenyls
For: cheapest plasticiser which converts rigid PVC (guttering) into flexible PVC (raincoats). The next cheapest would increase the price of the latter by perhaps 10 per cent.
Against: similar effects to DDT.

4 You are provided with a sample of about 10 g of a rather dark coloured solid and told that it is a newly discovered non-radioactive isotope of astatine, $^{220}_{85}At$. What tests would you apply to check this statement? Describe your actual procedures and expected observations.

5 Write balanced equations for
(a) the reaction of hydrogen bromide with manganese dioxide;
(b) the reaction of phosphorus trichloride with water;
(c) the reaction of sulphuric acid with potassium iodide;
(d) the reaction of acidified potassium chlorate with iron(II) salts;
(e) the reaction of ammonia with chlorine.
In each case calculate the mass of halogen or halogen-containing product obtainable from 100 g of halogen-containing reagent.

6 The following statements are not correct in all respects. There may be exceptions, ambiguities, omissions or deliberate mistakes. Write out a correct version of the statement and comment on the theoretical background to it.
(a) Silver halides are insoluble.
(b) The more electronegative is the element X, the stronger the acid HX.
(c) Halogens are most often found in oxidation states I and VII.
(d) Solubilities of compounds can be inferred from lattice energies.
(e) Halogens react with methane to form haloalkanes.

7 0.38 g of iodine react with 0.1 litre of chlorine to form orange crystals, no other product being obtained. These, when warmed gently formed a reddish liquid, which contained iodine and chlorine and weighed 0.485 g. This liquid was triturated (ground up) with potassium bromide, and after the excess liquid was removed by washing with dichloromethane, a red solid was left. On heating, coloured vapours were evolved. The solid remaining was soluble in water, and this solution with acidified silver nitrate gave a white precipitate soluble in dilute ammonia.
(a) Suggest formulae for the various compounds, and write appropriate equations.
(b) Account for the formation of the final potassium halide, as opposed to other possible halides (consider lattice energies).

9
The Noble Gases

A group with scientific ideas

9.1 DISCOVERY OF THE NOBLE GASES

The first noble gas to be discovered was helium, in 1868, although no one obtained a sample of the gas for some thirty years! In the 1860s the spectroscope had just been invented and the frequencies of light emitted by different elements vaporised in a flame were being catalogued. It was noticed that light from the sun was lacking in certain of these frequencies, so that fine dark lines appeared in the visible spectrum of the sun. It was then shown that the frequencies which an element produced when excited were the same as those absorbed when light passed through its vapour. On this basis, the elements present in the sun were identified; but one set of lines was unaccounted for. This suggested that the sun contained an element unknown on earth. The element was christened helium, (Greek *helios* = sun) but no terrestrial sample was obtained until about 1900.

The second noble gas to be discovered was argon. In 1895, Lord Rayleigh was determining, for reference purposes, the densities of all common gases to an accuracy of at least 1 in 1000. He showed that nitrogen prepared in chemical reactions had a density of 1.2505 gl^{-1} whereas 'nitrogen', the residual gas from air after removing oxygen (by Cu) and carbon dioxide (by KOH) had a density of 1.2575 gl^{-1}. Cavendish in 1785 had noted the same thing, but he either did not recognise the significance of his result or was unable to convince others of its significance. Ramsay, however, continued the work of Lord Rayleigh, and looked for an impurity in the 'atmospheric' nitrogen. He passed the sample over heated magnesium (to convert nitrogen to magnesium nitride, Mg_3N_2) and was left with the gas argon, density 1.783 gl^{-1}.

For a few years, there were difficulties about fitting just *one* extra element into the periodic table, next to potassium. This stimulated a search for other elements similar to argon. Neon, kryton and xenon were all isolated from the atmosphere shortly afterwards. Later helium was isolated from natural gas, and radon was obtained from the decomposition of radium. A complete new group was therefore added to the periodic table.

9.2 SUMMARY OF THE PROPERTIES AND USES OF NOBLE GASES

Element	*Atomic number* *Electronic structure* *Atomic mass*	*Atomic radius*/nm *1st ionisation energy*/kJmol^{-1}	*Parts per million in air* (by volume) *Solubility in water* (cm^3 per cm^3 H_2O) *Boiling pt.*/K	*Uses*
Helium	2 2: 4	0.12 2400	5 (about 0.5% in many natural gas sources) 0.010 4	Balloons (meteorology) Mixture with oxygen for divers' breathing (does not dissolve in body tissue as readily as does nitrogen under pressure) Liquid for superconducting magnets
Neon	10 2:8 20.2	0.16 2100	15 0.011 27	Neon lighting (advertising, and warning lamps)
Argon	18 2:8:8 40	0.19 1500	9300 0.058 87	Electric lighting (inert coolant): protection of reactive metals in processing (stainless steel and aluminium welding, titanium)
Krypton	36 2:8:18:8 83.8	0.20 1350	0.05 0.11 120	Electric light (inert coolant for high-intensity lamps)
Xenon	54 2:8:18:18:8 131.3	0.22 1200	0.006 166 0.24	Electric light (as krypton)
Radon	86 2:8:18:32:18:8 isotope 222 $t_{\frac{1}{2}} = 4$ days	– 1050	Nil – 202	Cancer treatment (radioactivity)

9.3 PROPERTIES OF THE NOBLE GASES

For seventy years the noble gases remained inert gases, no compounds being known. This stability was and still is associated with an outer shell containing eight electrons (He, two electrons). This is the well-known stable 'octet', which resists expansion to form covalent bonds or anions since extra electrons would have to be placed in an orbital of substantially higher energy (e.g. 3d if 3s and 3p are full or 3s if 2s and 2p are full). Such electrons can be held only weakly as there is no increase in the positive charge on the nucleus to help in their retention. On the other hand, removal of electrons from the noble gases is also difficult, since in moving across a period (e.g. Na–Ar),

(1) the positive charge on the nucleus has increased eight units,
(2) the radius of the isolated atom has decreased,
(3) therefore the outer electrons are more strongly held.

Cations and co-ordinate bonds are therefore unlikely: the first requires the outright loss of an electron, and the second the sharing of a pair. In accordance with this the noble gases are found to be monatomic, to have low boiling points, and to form very few compounds. Their uses are essentially those of inert diluents. Argon and neon are obtained directly from the fractional distillation of air, but

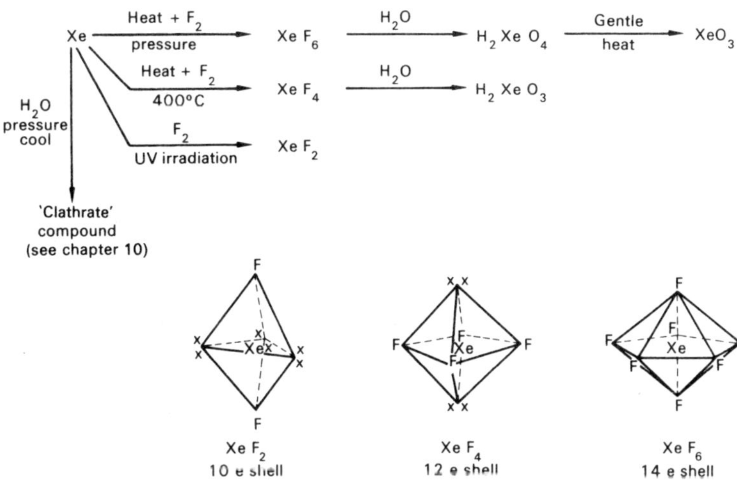

Fig. 9.1 Chemical properties of xenon.

krypton and xenon are found in the residue from the distillation of liquid oxygen. The remaining oxygen is then burnt off using hydrogen, leaving the krypton and xenon. Helium is obtained from natural gas, and radon from the radioactive decay of radium. Argon, forming 1 per cent of the atmosphere, is quite common (definitely *not* a rare gas) and cheap. The others are more expensive and are used for specialised purposes. Details are given in the table in section 9.2.

The 'stable octet' proved almost too successful an idea. Although some theorists noted the similarity of the ionisation potential of xenon (1170 $kJmol^{-1}$) to that of iodine (1010 $kJmol^{-1}$) and iodonium compounds such as $I(C_5H_5N)_2^+$ were known, no serious attempt was made to prepare noble gas compounds. In fact, the first such compound was discovered by accident, in an attempt to prepare $(SiO_2^+)(PtF_6^-)$. Silica and platinum(VI) fluoride were being heated in an inert atmosphere, xenon being used because the argon had run out. To everyone's astonishment, the crystals contained no oxygen or silicon and proved to be $Xe^+ PtF_6^-$. Once one compound had been made, the impossible became the everyday, as several laboratories heated xenon with fluorine in an attempt to make xenon fluorides; an experiment that might have been done at any time since the technology of handling fluorine had been solved (as early as 1944, in connection with the purification of uranium for the atomic bomb).

Several fluorine and oxygen compounds of xenon are now known, and also some of krypton (KrF_2, KrF_4). All of them are powerful fluorinating or oxidising agents, and the oxygen compounds in particular tend to be unstable.

QUESTIONS

1 Illustrate the tendency for scientists to ignore results which do not fit the existing and accepted 'scheme of things' by referring to
(a) Cavendish's discovery of argon,
(b) Bartlett's discovery of noble gas compounds.

2 Explain
(a) why xenon and krypton most readily form compounds;
(b) why helium is useful for mixing with oxygen in mixtures for divers to breathe;
(c) why 'atmospheric' nitrogen should be *heavier* than 'chemical' nitrogen;
(d) how it is possible to detect an element at a distance of 150 million km.

3 The following statements are not correct in all respects. There may be exceptions, ambiguities, omissions or deliberate mistakes. Write out a correct version of the statement and comment on the theoretical background to it.
(a) The noble gases, being unreactive, are not particularly useful.
(b) The noble gases are unreactive because their outer shell contains a stable octet of electrons.
(c) Helium, neon, argon, etc., are rare gases.
(d) Monatomic gases are ideal gases in the sense of the kinetic theory.

10
Hydrogen

A unique element on account of its size

10.1 INTRODUCTION

In many editions of the periodic table hydrogen is placed at the head of either group I or group VII. In reality, it bears little resemblance to either of these groups in anything other than valency. Its uniqueness arises from a combination of its low relative atomic mass, its small size, and the limitation of only two electrons in the first shell when hydrogen is involved in bonding.

Hydrogen is easily the most abundant element in the universe, and the radiation from main sequence stars relies on the overall reaction:

$$4\,{}^{1}_{1}H \longrightarrow {}^{4}_{2}He + 2\,{}^{0}_{1}e^{+}(\text{positron}) + \gamma\ (\text{radiation})$$

However, the free element is very scarce in the earth's atmosphere, not only because of the presence of oxygen but also due to the high velocity of hydrogen molecules. According to the kinetic theory, molecules of all gases at a given temperature have the same average kinetic energy. Since kinetic energy is measured by $\frac{1}{2}mv^2$, where m = mass of molecule and v = velocity, the low mass of the hydrogen molecule implies a very high speed, in fact about 1.7 km sec^{-1} at 0 °C. Oxygen, with a mass sixteen times as large, has, according to the expression above, an average velocity only one quarter that of hydrogen. Hydrogen molecules possess more than the escape velocity for a planet the size of the earth. Hence only planets as large as Saturn or Jupiter can retain hydrogen in their atmosphere. All the hydrogen which occurs on earth is combined, principally as water, but also as hydrocarbons and other organic compounds and as ammonium salts.

10.2 PROPERTIES RELATED TO SIZE

10.2.1 **Bond energies**

Because of the small size of the hydrogen atom, covalent bonds to hydrogen are very short, and hence very strong. Table 10.1 sets out some bond energies involving hydrogen, and compares them with similar bonds to carbon, an element of the same electronegativity.

The hydrogen molecule, of course, has the shortest and almost the strongest single bond of all; 0.07 nm and 440 kJmol^{-1}. Therefore, hydrogen and compounds containing it are generally unreactive—at room temperature! However, once the necessary activation energy has been supplied by heat, or light, the energy of the new bonds formed is sufficient to activate the rest, and there may

Bond	*Bond energy/* kJmol^{-1}	*Bond length/* nm	*Bond*	*Bond energy/* kJmol^{-1}	*Bond length/* nm
H–C	410	0.11	C–C	350	0.15
H–Cl	430	0.13	C–Cl	340	0.18
H–Br	370	0.14	C–Br	280	0.19
H–O	460	0.095	C–O	360	0.14
H–N	390	0.10	C–N	300	0. 15

Table 10.1 Selected bond energies and lengths.

be an explosion. For instance, both hydrogen and methane (CH_4) are quite unreactive towards oxygen at room temperature, and such gas mixtures are stable for periods of years. Apply a match, however, and there will be an immediate explosion. The temperature of the match flame is in the region of 500 °C. Similar considerations apply to the reaction of hydrogen with the halogens to form hydrogen halides, except that fluorine, having an exceptionally weak bond itself, will react spontaneously.

Bond energies for H–X bonds are set out in table 10.1. There is a periodic variation, associated with the covalent radius of the atom X.

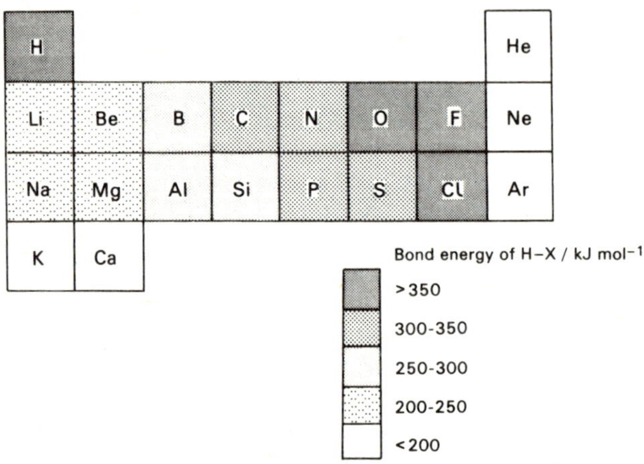

Fig. 10.1 Bond energies and periodicity.

10.2.2 Hydrogen bonding

Owing to its very small size, hydrogen can enter into 'hydrogen bonding'. When hydrogen (electronegativity value 2.0) is combined with nitrogen (3.0), oxygen (3.5), or fluorine (4.0), the covalent bond is very polar, with electrons of the bond spending most of their time away from the hydrogen atom. A dipole therefore develops which attracts neighbouring polar molecules. These cases are, however, exceptional. Because of the small size of the other atoms involved, it is possible for the two dipoles to approach very closely, thus sharply increasing the amount of attraction to the point where it justifies the special title of hydrogen bonding.

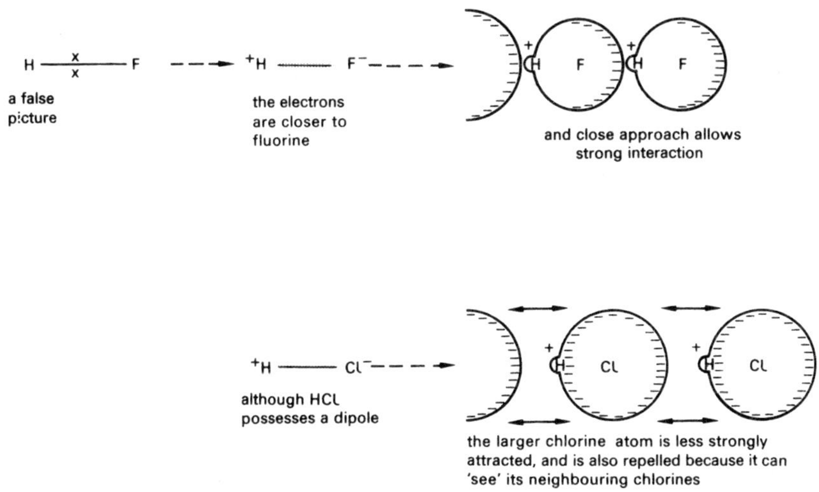

Fig. 10.2 Origin of hydrogen bonds.

Hydrogen bonding only occurs in compounds with N–H, O–H or F–H bonds (with the exception of a few special cases such as $CHCl_3$ and HCN), but does make a striking difference to the properties of such compounds. It has been used to explain the following properties:

(i) *Association.* If the boiling points of hydrides of various elements are plotted against the period of the periodic table in which the element falls, fig. 10.3 is obtained. Ammonia, water and hydrogen fluoride are clearly exceptional. Hydrogen bonding allows such molecules to associate (··H–F···H–F···H–F···) thus increasing the work needed to separate individual molecules into the vapour, and hence increasing the boiling point. Values of the molar heat of vaporisation also show the presence of association clearly.

If it were not for this effect, there would be no water on the earth's surface, and life as we know it—which depends on aqueous solutions—would not exist.

Another effect of hydrogen bonding is to cause the association of carboxylic

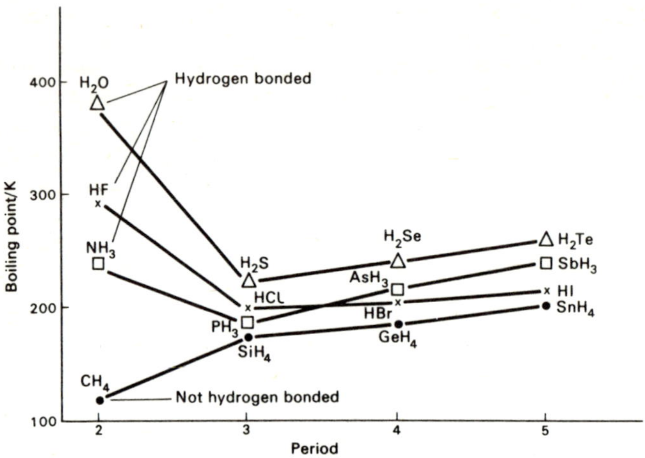

Fig. 10.3 Boiling points of hydrides.

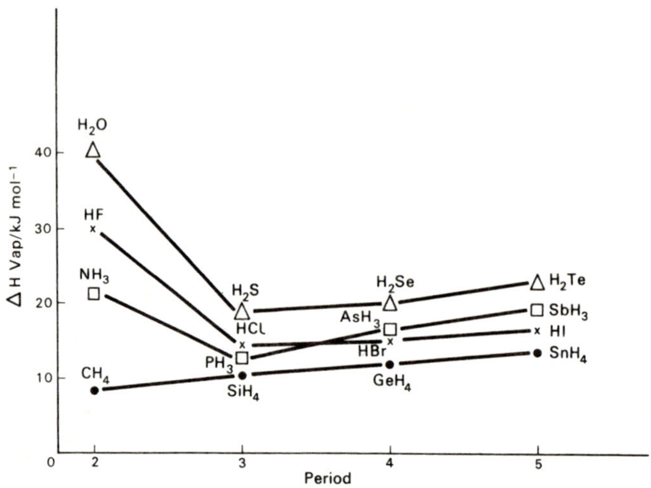

Fig. 10.4 Molar heats of vaporisation of hydrides.

acids. These exist as dimers in non-aqueous solvents, or in the vapour phase just above their boiling points.

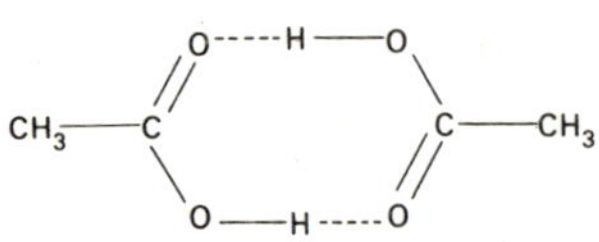

(ii) *Solubilities.* When a solute dissolves in a solvent, it must displace a number of solvent molecules. If the solvent molecules are tightly bound together—by hydrogen bonding, for instance—the solute must be capable of forming equally strong associations with the solvent. In particular, substances which dissolve readily in water must either be ionic—to attract the dipole of the water molecule—or else form hydrogen bonds with water. Some examples are given in table 10.2.

Substance	*Formula*	*Dipole*/D	H–*bond*	*Solubility*
Ethanol	$CH_3.CH_2.OH$	1.7	√	mixes completely
Chloroethane	$CH_3.CH_2.Cl$	2.0	×	insoluble
Ethanoic acid	$CH_3.CO.OH$	1.7	√	mixes completely
Methyl methanoate	$H.CO.OCH_3$	1.8	√	soluble

Table 10.2 Solubility and hydrogen bonding.

This effect is significant in biochemistry, since many foodstuffs contain hydrogen bonds, and are soluble (or readily hydrolysed to soluble substances) so that they can be transported in the blood.

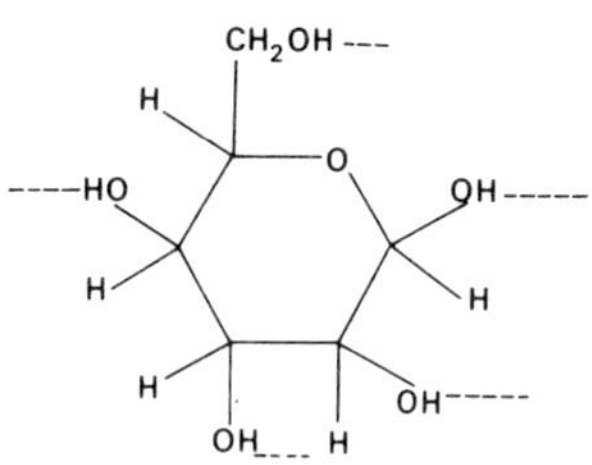

Glucose (hydrolysed from starch)

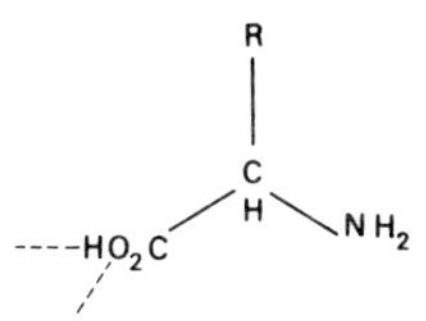

Amino acid (hydrolysed from

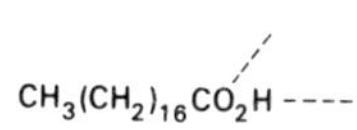

Fatty acid (hydrolysed from fat)

(iii) *Crystal structure.* Molecules which have strong hydrogen bonding often crystallise so that the crystal contains hydrogen bonds, often at the expense of closer packing. Ice is a notable example. The structure is shown in fig. 10.5.

The length and linearity of the O–H···O bonds hold the water molecules apart to such an extent that the density of ice falls below that of water. Ice therefore floats, fortunately for the fish, who can hibernate safely in the mud. Ammonium fluoride (similar to ice, with N and F replacing alternate O), urea and boric acid are other examples of crystals affected by hydrogen bonding.

Often such crystals contain molecular cavities, since close-packing is not observed. Foreign molecules can be incorporated in such spaces, without reaction, to form cage structures called clathrates. Argon, for instance, can be frozen into water under a pressure of eight atmospheres to form Ar.8 H_2O.

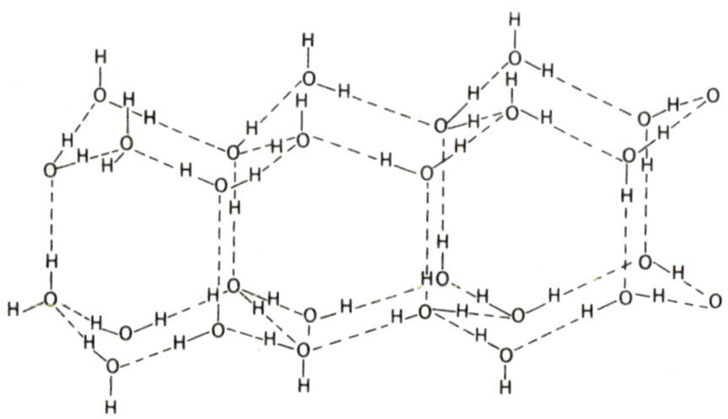

Fig. 10.5 Crystal structures affected by hydrogen bonding.

(iv) *Abnormal acidities.* There are a number of cases where the anion formed from an acid on dissociation can form a strong hydrogen bond. The acid therefore dissociates more readily. A simple example is hydrogen fluoride, which, unlike the other hydrogen halides is quite strongly acidic as the pure liquid.

$$3\,HF \rightleftharpoons (H{-}F\cdots H)^+ + (F\cdots H{-}F)^-$$

Other examples of unusually strong acids occur in organic chemistry.

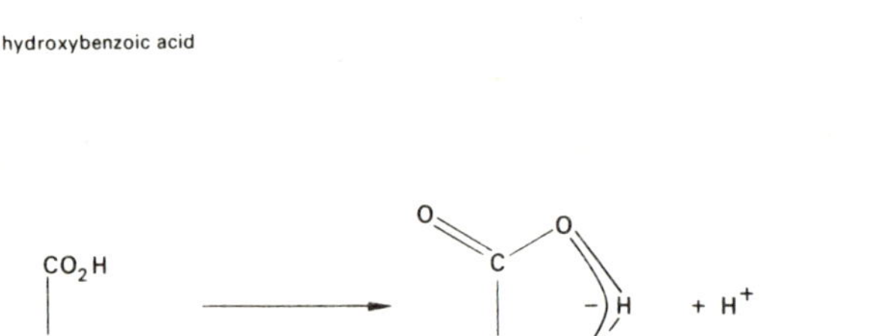

2- hydroxybenzoic acid

Oxalic acid

10.2.3 **Interstitial compounds**

Because of its small size, a hydrogen atom can fit into the holes in a close-packed metal lattice. There are two types of holes, one surrounded by four atoms (the tetrahedral hole) and one surrounded by six atoms (the octahedral hole). The latter is too big to hold the hydrogen, but is suitable for carbon or nitrogen atoms (see chapters 4 and 5). The former is too small for carbon or nitrogen, but holds a hydrogen atom. There are two tetrahedral holes for every metal atom (make a model to confirm this), and since any or all holes may be filled, interstitial hydrides typically have a formula varying between MH_0 and MH_2, depending on the pressure. The reversible nature of the absorption has prompted interest in these compounds as stores for hydrogen, should hydrogen ever become a fuel in widespread use as forecast by some authorities for the twenty-first century. One alloy of lanthanum and niobium will store, reversibly at atmospheric pressure, six hundred times its own volume of hydrogen—equivalent to a cylinder at 600 atmospheres pressure.

10.3 PROPERTIES DEPENDING ON MASS

Hydrogen has a melting point of −259 °C and a boiling point of −252 °C, lower than any substance except the noble gas helium. Its escape from the earth's atmosphere has already been referred to. The density of the solid is the lowest of any; indeed, there is more hydrogen in a cubic centimetre of water than in a cubic centimetre of solid hydrogen.

Hydrogen has two stable isotopes, normal hydrogen (99.99 per cent) containing one proton in the nucleus, and deuterium (0.01 per cent) containing one proton and one neutron. A third isotope, tritium, with two neutrons, is also known, but it is radioactive and does not occur naturally. The low mass of normal hydrogen implies that the percentage change in mass from one isotope to another is larger than for any other element. Differences in physical properties in substances containing deuterium are therefore very noticeable. For example, the boiling points of H_2O and D_2O are 100 °C and 101.8 °C respectively: their densities are 1.0 and 1.1. Rates of reaction are also affected; the oxidation of CH_3CHO is six times faster than that of CH_3CDO. The last effect is called *kinetic isotope effect*, and is sometimes used in tracing the mechanism of a reaction. Deuterium compounds are used generally as labels in tracing the fate of hydrogen atoms in a reaction.

Deuterium oxide is readily available from the large-scale electrolysis of water. D^+ ions are discharged more slowly than H^+ ions, so that the residue is enriched in deuterium. Repeated electrolysis of D_2O-enriched residues eventually leaves pure deuterium oxide. Other deuterium compounds are made from it by high-yield reactions avoiding waste. For example

$$D_2O \xrightarrow{SOCl_2} DCl$$

$$D_2O \xrightarrow{SO_3} D_2SO_4 \xrightarrow{NaNO_3} DNO_3$$

$$D_2O \xrightarrow{CaC_2} C_2D_2 \xrightarrow[D_2SO_4/Hg^{2+}]{D_2O} CD_3CDO$$

10.4 PROPERTIES DEPENDING ON ELECTRONIC CONFIGURATION AND SIZE

10.4.1 **General**

Hydrogen forms both positive and negative ions since its one electron in the outer shell may be lost (giving an empty outer shell) or an extra one may be gained (giving a full outer shell). However, H^+ ions, which consist of a bare proton, are not found in compounds. In solvents, the energy of solvation ensures that ions such as H_3O^+ (in water) or NH_4^+ (in liquid ammonia) are formed. In crystals, the enormous lattice energy which might be expected from such a small ion is never developed since the anions touch and repel each other before coming close to the proton. Therefore the ionisation energy of hydrogen cannot be recovered from the lattice energy, as happens for other cations.

Hydride ions, H^-, are unstable in most common solvents, either by a redox reaction

$$H^- + CHCl_3 \longrightarrow CH_2Cl_2 + Cl^-, \text{ etc.}$$

or by acting as a base

$$H^- + H_2O \longrightarrow H_2 + OH^-$$

Hydrides are most commonly encountered as solids.

The energies involved in the formation of the ions are as follows:

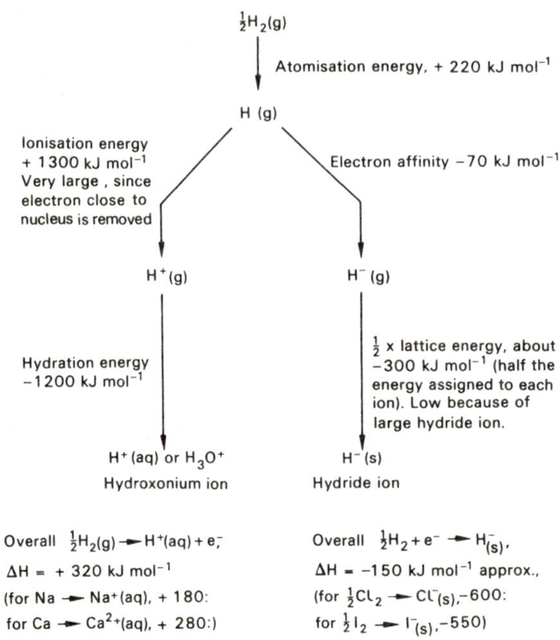

Fig. 10.6 Energies of formation of ions from hydrogen.

The endothermic formation of the familiar hydroxonium ion compared with the exothermic formation of the rare hydride ion looks unusual. Remember, though, that only half a cycle is shown. The 320 $kJmol^{-1}$ needed to form the hydroxonium, or any other similar ion, has to be paid for by the exothermic formation of the corresponding anion. The energy of formation of the hydroxonium ion is comparable with other cations, and it will form a comparable range of salts. However, the energy of formation of the hydride ion is substantially less than other anions, and the ion will only be formed if the cation needs little energy for its formation.

10.4.2 **The hydride ion**

As explained above, ionic hydrides are only found with easily formed cations. In practice, this includes group I and group II metals only. The ionic nature of these metal hydrides is shown by

(i) the crystal structure, which resembles the halides,
(ii) their electrolysis, in which hydrogen is liberated at the anode.

$$H^- \longrightarrow \tfrac{1}{2}H_2 + e^-$$

The electrolyses must be carried out in solution in molten group I chlorides because, except for lithium hydride, all decompose to their elements below their melting points.

Ionic hydrides react with water, as above, and are good reducing agents. They are sometimes spontaneously inflammable in moist air. Lithium aluminium hydride, $LiAlH_4$, is used extensively as a powerful reducing agent in organic chemistry.

The radius of the H^- ion is nominally as large as the iodide ion (0.21 nm), since the nucleus of one proton is unable to hold twice as many electrons strongly. However, the hydride ion is 'squashy' and its radius may decrease to as little as 0.13 nm in certain crystals.

10.4.3 **The hydroxonium ion**

A bare proton polarises water molecules so powerfully (see p. 36) that it forms bonds with one or more of them. Although usually written H_3O^+ or $H^+(aq)$, the principal cations in molar acids are $H_9O_4^+$ and $H_{11}O_5^+$. Acids will, of course, react with bases in a number of well-known reactions. For example:

carbonates ⟶ carbon dioxide	$2H^+ + CO_3^{2-} \longrightarrow H_2O + CO_2$
hydroxides ⟶ water	$H^+ + OH^- \longrightarrow H_2O$
sulphides ⟶ hydrogen sulphide	$2H^+ + S^{2-} \longrightarrow H_2S$

More examples are given in chapter 12, and for a general discussion of acid/base reactions, you are referred to a book on physical chemistry.

The formation of acids from covalent molecules can be represented as follows:

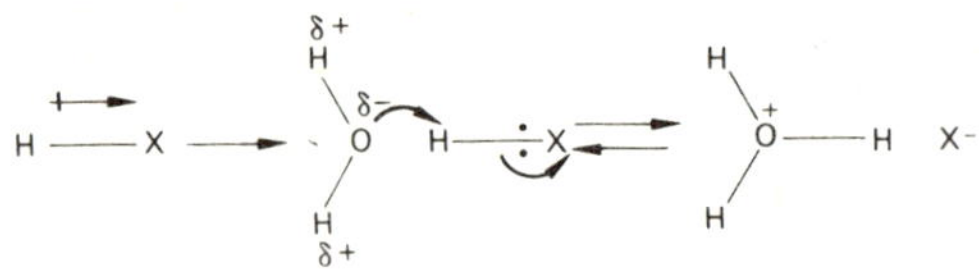

Dipolar molecule ... attracts dipole of water molecule ... there is competition for the proton, which may be transferred to the water

The proportion of molecules which transfer a proton to water reflect the strength of the acid. Two things favour a strong acid, therefore: a weak H–X bond, which is also strongly polar. Hydrogen, therefore, forms strong acids with large electronegative atoms, such as the halogens.

The polarity of the HX bond, and hence the strength of the acid, is increased in many cases by attaching electronegative atoms to X. Hence the order of acidity is $HClO_4 > HClO_2 > HClO$.

10.4.4 **Hydrides and periodicity**

For the reasons discussed above, elements on the left-hand side of the periodic table form salts with hydrogen, containing the hydride ion, H^-. Elements on the right-hand side, however, form covalent hydrides, which may dissociate to yield protons, H^+. The hydride ion, of course, reacts with water to form hydroxides. There is a weak periodicity in these and other properties, summarised in table 10.3. Fuller details of individual properties are given under the other elements concerned.

10.5 INDUSTRIAL CHEMISTRY OF HYDROGEN

Most of the hydrogen produced by the chemical industry in the U.K. arises from methane (natural gas). This is treated with steam at 700 °C when it is 'reformed' to produce carbon monoxide and hydrogen.

$$CH_4 + H_2O \longrightarrow CO + 3H_2$$

This mixture is passed with excess steam over an iron catalyst at 400 °C, when the carbon monoxide is largely oxidised to carbon dioxide.

$$CO + H_2O \longrightarrow CO_2 + H_2$$

The carbon dioxide is removed by absorption in potassium carbonate solution under pressure, which can later be reheated to regenerate the carbonate.

$$K_2CO_3 + H_2O + CO_2 \longrightarrow 2KHCO_3$$

Some hydrogen is also produced by a similar process using naphtha (a petroleum distillate, b.pt. about 40 °C) or by partial oxidation of heavier oils. This process has declined sharply in importance in the U.K. since North Sea gas came on stream, but is still widely used elsewhere. The water-gas process, in which steam was passed over hot coke, went out of use nearly twenty years ago, and is not likely to return while natural gas is available.

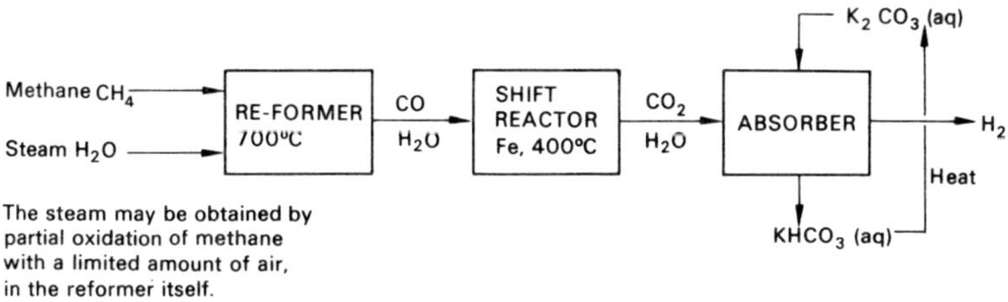

Fig. 10.7 Flowsheet for the industrial production of hydrogen.

The following reaction schemes sum up the uses to which hydrogen is put (for details see other chapters, or an organic chemistry book).

Hydrogen $\xrightarrow[\text{Fe, 450°C, 300 atm.}]{N_2}$ Ammonia (for fertilisers, nitric acid, explosives, dyes, etc.) 50% of production is taken up in this way

(U.K. 2×10^8. m^3 per year)

$\xrightarrow[\text{Cr, 300°C, 300 atm.}]{\text{CO}}$ Methanol (for solvent, methanal and plastics) 25% of production is taken up in this way

$\xrightarrow[\text{Ni, 100°C}]{\text{oils (C=C)}}$ Fats (margarine, $-CH_2-CH_2-$)

$\xrightarrow[\text{Ni, 100°C}]{\text{Benzene } (C_6H_6)}$ Cyclohexane, C_6H_{12} (for nylon) and various other uses.

10.6 SUMMARY OF HYDROGEN CHEMISTRY

For industrial processes, see above: these reactions in the summary are essentially laboratory reactions.

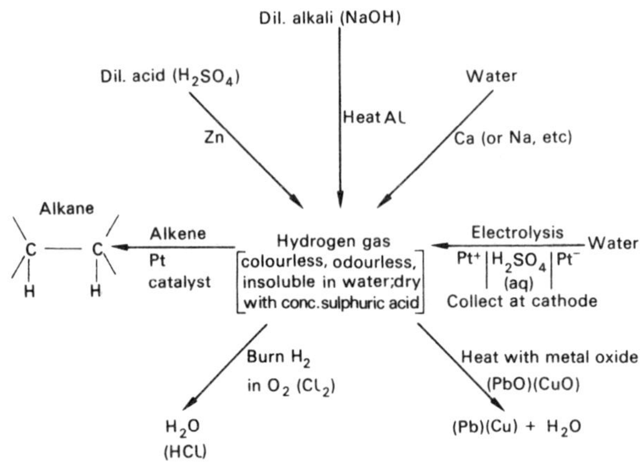

All equations are trivial except

$$\begin{cases} Al + 6OH^- \rightarrow Al(OH)_6^{3-} + 3e^- \quad \text{hexahydroxyaluminate (III)} \\ e^- + H_2O \rightarrow \tfrac{1}{2}H_2 + OH^- \end{cases}$$

Fig. 10.8 Summary of hydrogen chemistry.

10.7 SUMMARY OF ACID CHEMISTRY

Representative examples only are given.

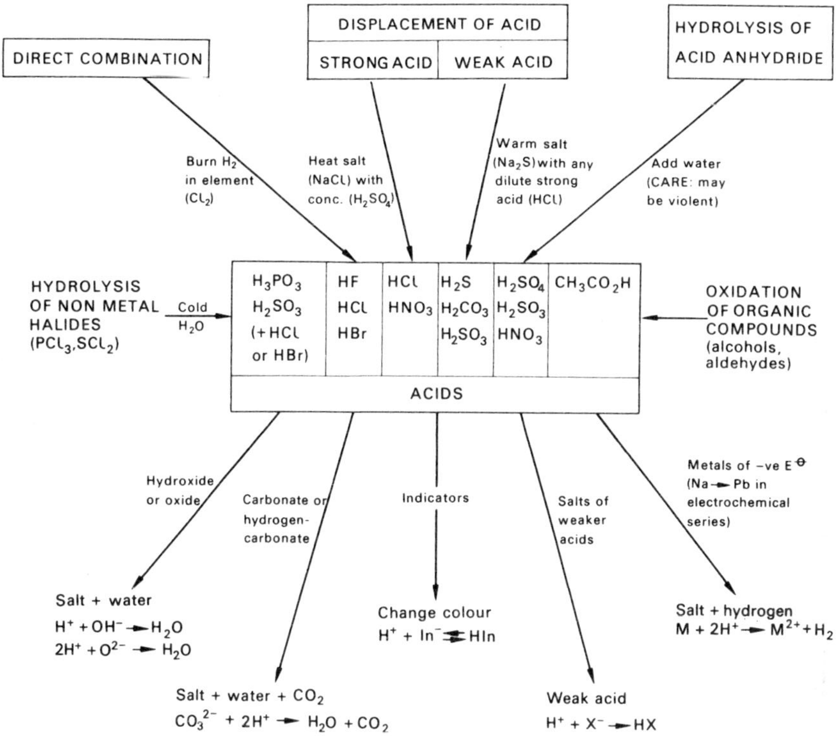

Fig. 10.9 Summary of acid chemistry.

Group	*Element*	*Properties of hydride*			
		Bonding	*State at s.t.p.*	*Reaction with water*	*Reaction with air*
I	Li	Ionic H^-	Solid	Violent, OH^- formed	Burns vigorously
II	Be	Covalent polymer	Solid	Not known	Burns vigorously
III	B	Covalent polymer	Gas	Neutral	Burns
IV	C	Covalent	Gas	Neutral	Burns
V	N	Covalent	Gas	Weakly basic	Burns only if $[O_2] > 30\%$
VI	O	Covalent	Liquid	Neutral	–
VII	F	Covalent	Liquid	Acidic	No reaction
I	Na	Ionic H^-	Solid	Violent, OH^- formed	Burns vigorously
II	Ca	Probably ionic H^-			Burns vigorously
III	Al	Not known	Solid	Not known	Not known
IV	Si	Covalent	Gas	Neutral	Ignites spontaneously
V	P	Covalent	Gas	Neutral	Ignites spontaneously
VI	S	Covalent	Gas	Weakly acidic	Burns
VII	Cl	Covalent	Gas	Acidic	Oxidised catalytically

Table 10.3 Periodicity and hydrides.

QUESTIONS

1 In some versions of the periodic table hydrogen is placed at the head of group I, in others at the head of group VII. In what respects, and why, does hydrogen resemble members of these groups? Is either of these classifications useful, or is it better to leave hydrogen on its own?

2 Write a systematic account of the preparations and reactions of acids.

3 Show how the stated principle is connected to the given phenomenon.

Principle	*Phenomenon*
(a) Hydrogen forms strong bonds	A mixture of hydrogen and oxygen explodes on ignition
(b) The formation of hydrogen bonds	Argon froms a hydrate $Ar.8\,H_2O$.
(c) The co-ordinate bond	Acids demonstrate their typical properties only in dilute solution
(d) Some acids are stronger than others	A test for carbonates is effervescence in the presence of dilute acid
(e) The number of neutrons in a nucleus is not directly related to the number of protons	Heavy water boils at 102 °C

4 Explain as if to someone with a smattering of science how the following properties depend on the size and structure of the hydrogen atom, and what the consequences would be for everyday life if the property did not hold, or was reversed.
(a) Hydrogen is absent from the earth's atmosphere.
(b) Ponds freeze from the top downwards in winter.
(c) Natural gas does not ignite in air below about 500 °C.
(d) Vinegar, soda water and lemon juice all have a bitter taste.
(e) Gin mixes with tonic water.
Also
(f) Calculate the volume of hydrogen obtainable from 1 cm^3 of each of the following:
H_2O; $LiH(\rho = 0.8\,g\,cm^{-3})$; $(CH_2)_n$ (polythene, $\rho = 0.8\,g\,cm^{-3}$); TiH_2 (interstitial, $\rho = 8\,g\,cm^{-3}$); solid H_2 ($\rho = 0.1\,g\,cm^{-3}$); solid CH_4 ($\rho = 0.7\,g\,cm^{-3}$). Review their uses as possible sources of hydrogen for fuel.

5 The following statements are not correct in all respects. There may be exceptions, ambiguities, omissions or deliberate mistakes. Write out a correct version of the statement and comment on the theoretical background to it.
(a) Because $\frac{1}{2}H_2 \longrightarrow H^-$ is more exothermic than $\frac{1}{2}H_2 \longrightarrow H^+$, hydrogen has a strong tendency to form hydrides.
(b) Non-metal hydrides are acids.
(c) The boiling points of hydrides increase down a given group in the periodic table.
(d) The size of an anion is related to the total number of electrons that it contains.
(e) Solutions of acids in water typically contain H_3O^+.

11
The Transition Metals

Elements possessing unfilled orbitals of low energy

11.1 INTRODUCTION

The key to understanding the properties of the transition metals is a familiarity with the relative positions of the energy levels available for their electrons.

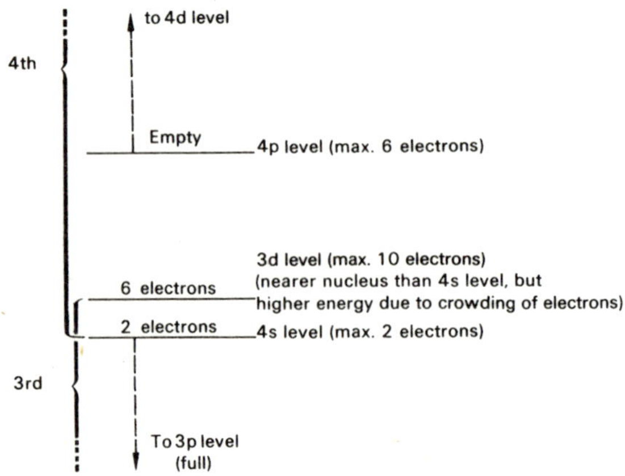

Fig. 11.1 Ground state of a typical transition element (iron, atomic number 26).

The third and fourth principal quantum shells are closer in energy than preceding shells, with the result that the highest energies in the third level overlap the lowest in the fourth. All the elements considered in this chapter, from scandium (atomic number 21) to zinc (atomic number 30), have their 4s level occupied—and since this level is in fact further from the nucleus than the 3d level, all present a more or less similar face to the world. Similarities may therefore be expected along the period, as well as down the group.

Once the 4s level is full, electrons enter the 3d orbitals. As long as either 3d or 4p orbitals remain unfilled, additional electrons can be accepted from other molecules or anions to form the complex ions characteristic of transition metals. Conversely, electrons may be lost, but since there are up to twelve electrons of similar energy (ten in the 3d orbitals and two in the 4s orbitals) the actual number involved is variable. The formal definition of a transition metal is: an element whose atom or ion possesses a partially filled d sub-shell. This excludes zinc, which is, however, dealt with here on account of its position in the periodic table and also as a contrast with those elements possessing unfilled d orbitals.

Element	*Electronic structure of atom* $1s^2\ 2s^2\ 2p^6\ 3s^2\ 3p^6$ plus		*Vacant orbitals*	*Typical ion*	*Number of electrons in the* 3d *and* 4s *level in the ion*
	3d	4s			
21 Scandium (Sc)	1	2	$3d^9$, $4p^6$	Sc^{3+}	0
22 Titanium (Ti)	2	2	$3d^8$, $4p^6$	Ti^{3+}	1
23 Vanadium (V)	3	2	$3d^7$, $4p^6$	V^{3+}	2
24 Chromium (Cr)	5	1	$3d^5$, $4s^1$, $4p^6$	Cr^{3+}	3
25 Manganese (Mn)	5	2	$3d^5$, $4p^6$	Mn^{2+}	5
26 Iron (Fe)	6	2	$3d^4$, $4p^6$	Fe^{2+}	6
27 Cobalt (Co)	7	2	$3d^3$, $4p^6$	Co^2	7
28 Nickel (Ni)	8	2	$3d^2$, $4p^6$	Ni^{2+}	8
29 Copper (Cu)	10	1	$4s^1$, $4p^6$	Cu^{2+}	9
30 Zinc (Zn)	10	2	$4p^6$	Zn^{2+}	10

Table 11.1 Electronic structure of transition elements.

Chromium and copper have anomalous electronic structures. The 3d and 4s orbitals are so close in energy that minor factors—in this case the extra stability of a filled or half-filled shell—can alter the electron distribution. The closeness of energy levels is also responsible for the loss of electrons from the 4s shell first when an ion is formed.

$$Zn(\equiv Ar + 3d^{10}4s^2) \longrightarrow Zn^{2+}(\equiv Ar + 3d^{10})$$
$$Fe(\equiv Ar + 3d^6 4s^2) \longrightarrow Fe^{3+}(\equiv Ar + 3d^5)$$

11.2 PROPERTIES OF TRANSITION METALS RELATED TO UNFILLED ORBITALS

11.2.1 **Physical properties**

In the solid state, metal atoms are held together by characteristic 'metallic bonds'. These are formed between atoms of elements which have a large number of unfilled orbitals of similar energy to the highest filled ones. Unfilled orbitals on neighbouring atoms can interact to form an extended and delocalised series of orbitals extending through the crystal lattice. Electrons promoted into this delocalised orbital can be transferred rapidly through the material. Hence metals have high thermal and electrical conductivity. This is especially true for transition metals because of the large number of electrons and unfilled orbitals involved. Copper and silver are notably good conductors.

In this type of bonding, the delocalisation appears to give a lattice of metal ions, completely permeated by a fluid made up of electrons. Bonding in the lattice is therefore non-directional, and this allows high co-ordination numbers to be achieved. In fact, the arrangement of atoms in a crystal of a transition metal is generally hexagonal close-packed, in which each atom has twelve nearest neighbours. Other arrangements are shown in fig. 11.2.

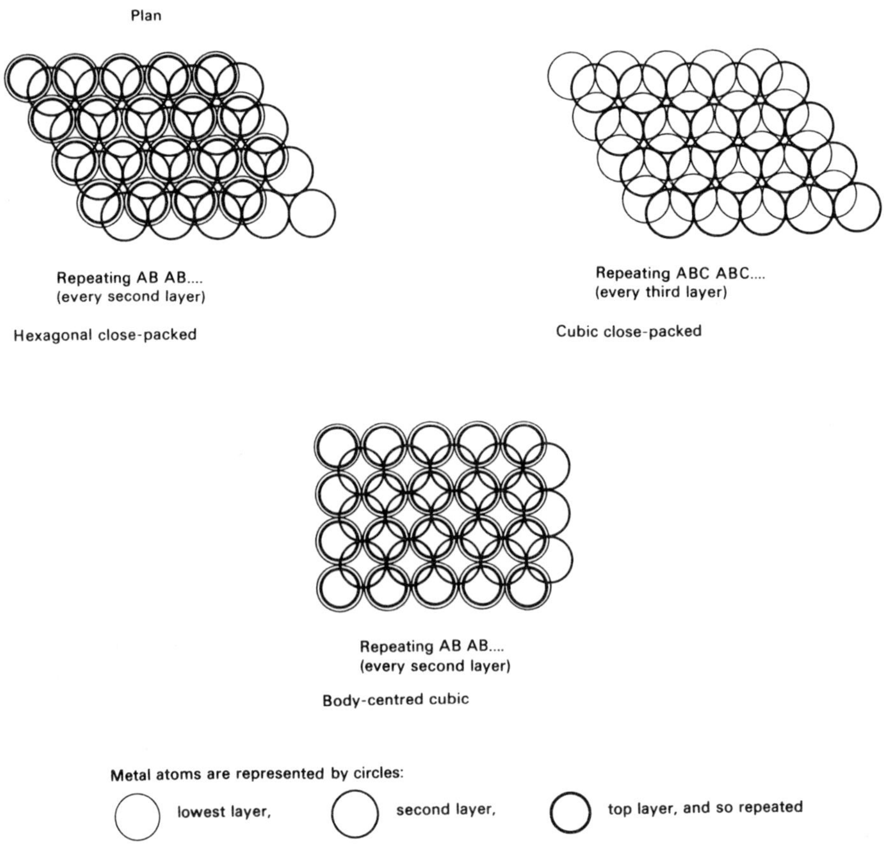

Fig. 11.2 Crystal arrangements.

Transition metals possess a low atomic radius, because the 3d sub-shell being filled is nearer the nucleus than the outermost 4s sub-shell which already contains electrons. This, coupled with the high co-ordination, results in many short, strong links between one atom and another. Consequently the density, hardness and melting point of transition metals are all high compared to metals in groups I and II (table 11.2).

The similar size of the atoms of these elements allows the ready replacement in a crystal lattice of atoms of one element by atoms of another element. Hence transition metals form a wide variety of alloys.

Element	Sc	Ti	V	Cr	Mn	Fe	Co	Ni	Cu	Zn	Na
Crystal structure	hcp or ccp	hcp or bcc	bcc	bcc	irreg. cubic	ccp or bcc	ccp or hcp	ccp or hcp	ccp	irreg. hcp	bcc
Atomic radius/nm	0.16	0.145	0.13	0.125	0.13	0.125	0.125	0.125	0.125	0.135	0.19
Density/ gcm^{-3}	3	4.5	6	7	7	8	9	9	9	7	1
Hardness (Moh)		6	6	7	5	5	5	5	3	2.5	0.5
Melting point/°C	1400	1700	1700	1900	1250	1550	1500	1450	1100	450	100

hcp = hexagonal, ccp = cubic-close packed (co-ordination 12)
bcc = body-centred cubic (co-ordination 8)

Table 11.2 Physical properties of transition elements.

Interstitial compounds (chapter 10) are also frequent and characteristic, the size of the spaces in the lattices being sufficient to accommodate the small atoms of carbon, hydrogen and nitrogen.

11.2.2 **Variable oxidation state**

There are many electrons of similar energy in the 3d and 4s levels. In general, the ionisation energy required to strip both shells is prohibitively high for the formation of a stable compound. On the other hand, there is no definite point at which further ionisation becomes clearly unprofitable (as it does, for example, at K^+ and Ca^{2+}). Nearly all transition metals, therefore, form compounds in several oxidation states. The exceptions are scandium (Sc^{3+} only, $4s^2$ and $3d^1$ completely stripped) and zinc (Zn^{2+} only, $4s^2$ stripped, leaving $3d^{10}$).

The range of oxidation states is extended by electron donation from ligands (see table 11.3), and a comprehensive list is given below.

One or two generalisations about oxidation numbers emerge:

(1) the common oxidation states usually include +2 or +3 or both;
(2) the maximum oxidation state is equal to the total number of 3d + 4s electrons as far as Mn, and thereafter decreases;
(3) the higher oxidation states occur in compounds with oxygen or fluorine (see chapter 8);
(4) the higher oxides are acidic, whereas the lower oxides are amphoteric or basic (see chapter 7);
(5) the lowest oxidation states often reduce water to hydrogen, and the highest oxidise it to oxygen;
(6) the highest oxidation states never occur as simple ionic compounds, the charge/radius ratio being too high for these small, highly charged ions to form

	Sc	Ti	V	Cr	Mn	Fe	Co	Ni	Cu	Zn
Lowest known oxidation state	+ 3	−1	−1	−2	−3	−2	−1	−1	+1	2+
0		✓	✓	✓	✓	✓	✓	✓		
+ 1			✓	✓	✓	✓	✓	✓	✓	
+ 2		✓	✓	✓	✓✓	✓✓	✓✓✓	✓✓✓	✓✓✓	✓✓✓
+ 3	✓✓✓	✓✓	✓✓	✓✓✓	✓✓	✓✓✓	✓✓	✓	✓	
+ 4		✓✓✓	✓✓	✓	✓✓✓	✓	✓	✓		
+ 5			✓✓✓	✓✓	✓	✓				
+ 6				✓✓	✓✓	✓				
+ 7					✓					
Highest known oxidation state	+ 3	+ 4 oxide halides	+ 5 oxide fluoride	+ 6 oxide fluoride	+ 7 oxide $KMnO_4$	+ 6 FeO_4^{2-}	+ 4 CoF_6^{2-}	+ 4 NiF_6^{2-}	+ 3 CuO_2^{-} CuF_6^{3-}	+ 2

Oxidation states corresponding to 3d shell ($3d^{10}$)

Oxidation states corresponding to half filled 3d shell ($3d^5$)

Oxidation states corresponding to vacant 3d shell

Good reducing agents

Stable in atmospheric conditions

Good oxidising agents

✓ = Known ✓✓ = Stable in aqueous solution ✓✓✓ = Stable in geological time

Table 11.3 Oxidation states of transition elements.

(see chapter 4). Covalent molecules or complex ions are formed instead. Thus, oxides and fluorides such as TiO_2 and VF_5 are covalent: Cr(VI) and Mn(VII) are found as oxy-anions CrO_4^{2-} and MnO_4^- respectively.

Oxidation states of zero may seem unusual. The uncombined element is of course in a state of zero. If it reacts with a neutral ligand—frequently carbon monoxide—it remains in that state.

$$Ni + 4CO \xrightarrow{100°C} \underset{\text{tetracarbonylnickel(0)}}{Ni(CO)_4}$$

The carbon monoxide can be replaced by other ligands, e.g. phosphorus trifluoride.

11.2.3 **Catalytic activity**

Variable oxidation states allow transition metal ions to alternately accept and reject electrons thus cycling between higher and lower oxidation states. In this way, they may help to transfer electrons from reductant to oxidant and catalyse redox reactions. For example, in the Deacon process, copper(II) oxide is used to convert hydrogen chloride (a waste product from organic chlorination) back to chlorine. The probable course of the reaction is

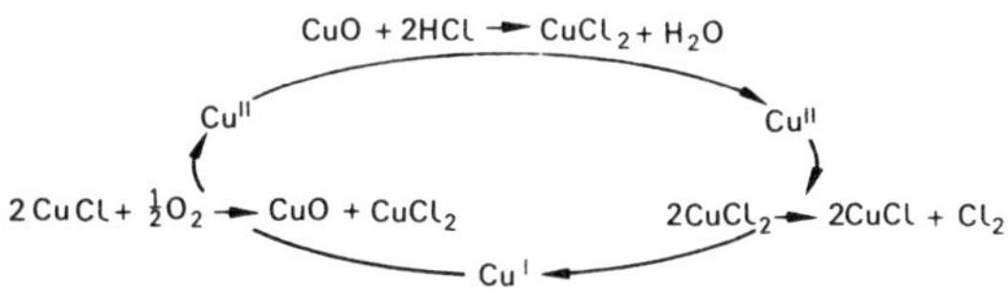

Other examples of catalysis involve the formation of intermediate complexes (cf. 11.3.1), particularly with alkenes as ligands. A limited list is given below, but this is merely a selection from the numerous processes catalysed by transition metals or their compounds.

Element	*Used as*	*Reaction catalysed*
Sc		No example: no variable oxidation state
Ti	$AlCl_3/Ti(C_2H_5)_4$	Polymerisation of ethene, propene
V	V_2O_5	$2SO_2 + O_2 \longrightarrow 2SO_3$ (contact process)
Cr	Cr_2O_3	$C_2H_5OH \longrightarrow CH_3CHO + H_2$
Mn	MnO_2	$H_2O_2 \longrightarrow H_2O + \frac{1}{2}O_2$
Fe	Fe_2O_3	$N_2 + 3H_2 \longrightarrow 2NH_3$ (Haber process)
Co	metal	$RCH{=}CH_2 + CO + H_2 \longrightarrow RCH_2CH_2CHO$ (oxo process)
Ni	metal	$RCH{=}CH_2 + H_2 \longrightarrow RCH_2CH_3$ (margarine)
Cu	CuO	$4HCl + O_2 \longrightarrow 2H_2O + Cl_2$ (Deacon process)
Zn	$ZnCl_2$	$ROH + HCl \longrightarrow RCl + H_2O$

Table 11.4 Examples of catalysis.

11.3 LIGANDS AND COMPLEX FORMATION

For a note on names, see chapter 1.

11.3.1 Complex ions

Transition metal ions rarely occur as the simple metal ion. Instead, complex ions are formed in which the simple ion is bound to two or more 'ligands'. Ligands are molecules or ions with non-bonding pairs of electrons which are initially attracted by the positive charge on the ion, forming permanent bonds by feeding these electrons into the vacant 3d and/or 4s orbitals of the transition metal ion. The ligands are then said to be co-ordinated.

The arrangement of ion and ligand was first proposed by Werner early this century. It brought order to a tremendously complex situation. For example, five compounds containing platinum(IV) chloride and ammonia are known.

Empirical formula	*Total number of ions* (deduced from conductivity)	*Number of chloride ions* (deduced from AgCl ppt.)	*Werner's formula*
$PtCl_4.6NH_3$	5	4	$(Pt(NH_3)_6)^{4+}4Cl^-$
$PtCl_4.5NH_3$	4	3	$(Pt(NH_3)_5Cl)^{3+}3Cl^-$
$PtCl_4.4NH_3$	3	2	$(Pt(NH_3)_4Cl_2)^{2+}2Cl^-$
$PtCl_4.3NH_3$	2	1	$(Pt(NH_3)_3Cl_3)^+Cl^-$
$PtCl_4.2NH_3$	0	0	$(Pt(NH_3)_2Cl_4)$ not charged

Table 11.5 Ammines of platinum(IV) chloride.

Werner deduced that besides the oxidation state—in this case IV—there was a characteristic number of ligands associated with each metal ion, called the co-ordination number. In the case of the ammines of platinum(IV) chloride, replacement is easy and any number of chloride ions and ammonia molecules to a total of six can make up the complex ion.

Most of the chemistry of transition metals, including all their aqueous chemistry, involves complex ions. Thus, when an uncomplexed ion is dissolved in water, the hydrated ion is formed

$$\underset{\text{(anhydrous, white)}}{(Cu^{2+}SO_4^{2-})} \longrightarrow \underset{\text{(blue hexaaquacopper(II))}}{Cu(H_2O)_6^{2+}} + SO_4^{2-}$$

The ligands may be replaced, partly or completely

$$Cu(H_2O)_6^{2+} + \underset{\text{(ammonia solution)}}{4NH_3} \longrightarrow \underset{\text{(diaquatetraammino-copper(III))}}{Cu(H_2O)_2(NH_3)_4^{2+}} + 4H_2O$$

$$\underset{\text{(green)}}{Ni(H_2O)_6^{2+}} + 6NH_3 \longrightarrow \underset{\text{(purple hexaamminonickel(II))}}{Ni(NH_3)_6^{2+}} + 6H_2O$$

The ligands themselves may be neutral (H_2O, NH_3) or negatively charged (Cl^- in $CoCl_4^{2-}$, tetrachlorocobaltate(II); CN^- in $Fe(CN)_6^{3-}$, hexacyanoferrate(III)). The complex formed may be positive, neutral or negative. The most common co-ordination number is six, but examples of two ($Ag(NH_3)_3^+$, diamminosilver (I)) and four ($CoCl_4^{2-}$) are not uncommon. In particular, lower co-ordination numbers are found in high oxidation states. (MnO_4^-, tetraoxomanganate(VII), CrO_4^{2-}, tetraoxochromate(VI)).

In the same way that the polarising power of Al^{3+} makes the dissociation

$$Al(H_2O)_6^{3+} \longrightarrow (Al(H_2O)_5OH)^{2+} + H^+(aq)$$

easy (chapter 4), so most of the M^{3+} ions in the first transition metal series form acidic solutions. Iron(III), chromium(III) and even zinc(II), are all examples. In higher oxidation states, the polarising power is increased so much that the dissociation of protons is complete. $Cr(H_2O)_6^{6+}$ and $Mn(H_2O)_6^{7+}$ do not exist: loss of protons and water molecules gives CrO_4^{2-} and MnO_4^-, corresponding to the strong acids H_2CrO_4 and $HMnO_4$ respectively.

11.3.2 **Colour**

Uncomplexed transition metal ions are usually colourless (e.g. anhydrous copper (II) sulphate). The interaction of ligand and ion is responsible for the colour so characteristic of transition metals. The colour arises because the d orbitals containing the electrons are not symmetrically arranged, some being closer to a ligand than others. Electrons in those close to a ligand are repelled, and such an orbital is of slightly higher energy. The energy required to promote an electron from an unaffected orbital to an affected one corresponds to a quantum of

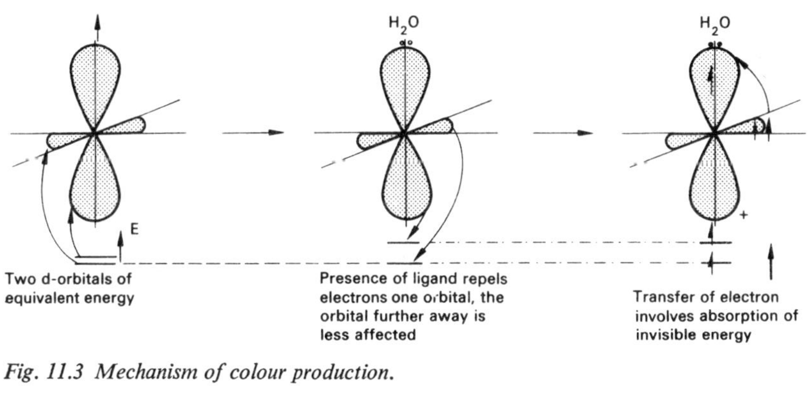

Fig. 11.3 Mechanism of colour production.

visible radiation. If the frequency used to promote the electron is in the red end of the spectrum, blue light remains and the substance appears blue, and vice versa.

Each transition metal ion once complexed has a characteristic colour. This colour usually undergoes only slight variation when the ligand alters:

$$\underset{\text{light blue}}{Cu(H_2O)_6^{2+}} \xrightarrow{NH_3} \underset{\text{dark blue}}{Cu(NH_3)_4(H_2O)_2^{2+}}$$

but when the number of ligands (and hence their position) alters, different orbitals are affected, and the colour changes:

$$\underset{\text{light blue}}{Cu(H_2O)_6^{2+}} \xrightarrow{HCl} \underset{\text{green}}{CuCl_4^{2-}}$$

Similarly, different oxidation states have different numbers of d electrons, and the colour again alters. For instance, if one *small* crystal of manganate(VII), which is purple, is added to a boiling solution of 50 per cent potassium hydroxide, it is reduced successively through all the oxidation states of manganese: Mn(VI) (green) $\longrightarrow$ Mn(V) (blue) $\longrightarrow$ Mn(IV) (brown) $\longrightarrow$ Mn(III) (green) $\longrightarrow$ Mn(II) (faint pink). The electrons for the reduction are provided by the hydroxide

$$2OH^- \rightleftharpoons 2e^- + H_2O_2 \longrightarrow H_2O + \tfrac{1}{2}O_2$$

Suitable precautions *must* be taken to guard against splashes in this experiment.

Common ion	*Higher oxidation state*	*Other cases*
Sc^{3+} colourless (no d-electrons to make transitions)		
Ti^{3+} green/violet	Ti^{4+} colourless (cf. Sc^{3+})	
V^{2+} violet	V^{3+} green	$V^{4+}(VO^{2+})$ blue $V^{5+}(VO_4^{3-})$ colourless
Cr^{2+} sky-blue	Cr^{3+} green/violet	$Cr^{6+}(CrO_4^{2-})$ yellow
Mn^{2+} pink		$Mn^{6+}(MnO_4^{2-})$ green $Mn^{7+}(MnO_4^-)$ purple
Fe^{2+} green	Fe^{3+} yellow/red	
Co^{2+} pink	Co^{3+} yellow/violet	(Co^{2+}, 4 ligands, blue)
Ni^{2+} green/purple		
Cu^{2+} blue		(Cu^{2+}, 4 ligands, green)
Zn^{2+} colourless (all d orbitals full, therefore no transitions)		

All ions have a co-ordination number of six (e.g. $M(H_2O)_6^{2+}$) unless stated

Table 11.6 Colours of transition metal ions.

11.3.3 Stereochemistry

Ligands are distributed in most cases on the repulsion principle. Six ligands, as far apart from each other as possible, occupy the corners of an octahedron.

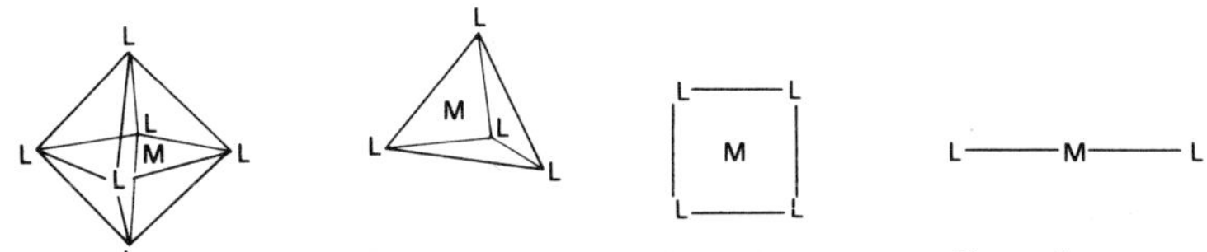

Hexaco-ordinate [all $M(H_2O)_6^{2+}$]

Tetraco-ordinate [usually oxo and chloro complexes, e.g. MnO_4^-, $CoCl_4^{2-}$, $CuCl_4^{2-}$]

Alternative tetraco-ordinate for Cu^{II}, Ni^{II}

Dico-ordinate $Ag(NH_3)_2^+$

Fig. 11.4 Arrangements of ligands.

Four ligands are usually to be found at the corners of a tetrahedron. However, nickel(II) and copper(II) usually form square tetra co-ordinate complexes, with the exception of $CuCl_4^{2-}$. Even octahedral complexes of copper(II) are distorted towards a square shape, two opposite ligands being further from the ion than the other four which form a square.

Isomers are also possible. Although beyond the scope of this book, a pleasant and testing half-hour may be spent investigating, with models, the isomeric possibilities of $(Co^{III}(NH_3)_2(CH_3NH_2)_2Cl_2)^+$, i.e. $(Co\ a_2b_2c_2)^+$, and others.

11.3.4 Redox potential

The standard electrode potential or redox potential refers to the hydrated ions in molar solution—$V(H_2O)_6^{3+}$, $Fe(H_2O)_6^{2+}$ and so on. Tabulated values of these potentials serve as useful guides to the aqueous chemistry of transition metals. In general, the redox system with the more positive potential can take electrons from the redox system with the more negative potential.

For the reaction $H^+ + e^- \longrightarrow \frac{1}{2}H_2$, $E^\ominus$ is defined as 0.0 volts. This is positive with respect to any redox couple having a negative potential, and as fig. 11.5 shows, this includes all the reactions $M^{2+} + 2e^- \longrightarrow M$ except for M = copper. Therefore, all of these metals except copper dissolve in molar acid to give hydrogen. Indeed, since the next potential, $M^{3+} + e^- \longrightarrow M^{2+}$, for Ti, V, and Cr is also below that of hydrogen, the divalent ions react further, forming trivalent ions and reducing the acid to hydrogen. The overall reaction for chromium is

$$Cr + 2H^+ \longrightarrow Cr^{2+} + H_2$$
$$Cr^{2+} + H^+ \longrightarrow Cr^{3+} + \tfrac{1}{2}H_2$$

However, the reaction with titanium and vanadium is very slow, and almost unnoticeable unless hot acid and finely divided metal are used. Redox potentials only indicate which way a reaction will go if it goes at all.

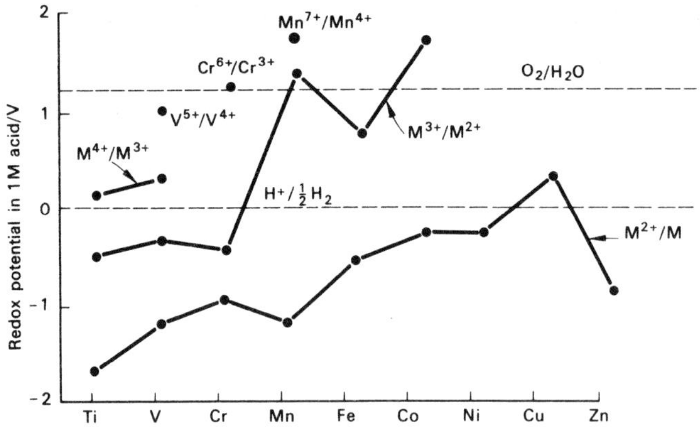

Fig. 11.5 Redox potentials of transition metals.

The potential $\frac{1}{2}O_2 + 2H^+ + 2e^- \longrightarrow H_2O$ $E^\ominus = 1.23\,V$ applies to acids in the presence of air, under normal atmospheric conditions. Notice that $Cu^{2+} + 2e^- \longrightarrow Cu$ $E^\ominus = 0.34\,V$ lies below this, so that copper tarnishes in rain water ($H_2O/CO_2/O_2$). $Fe^{3+} + e^- \longrightarrow Fe^{2+}$ can also be accomplished in air, so that solutions of iron(II) salts are unstable; and generally iron(II) salts are not kept as solutions.

The reduction of higher oxidation states involves protonation as well as reduction (see section 11.3.1)

$$MnO_4^- + 8H^+ + 2H_2O \longrightarrow \text{'}Mn(H_2O)_6{}^{7+}\text{'} \xrightarrow{5e^-} Mn(H_2O)_6{}^{2+}$$

Overall $$MnO_4^- + 8H^+ + 5e^- \longrightarrow Mn^{2+} + 4H_2O$$

Their reactions are therefore sensitive to acid.

$$MnO_4^- \longrightarrow MnO_2 \text{ (1 M acid)} \quad E^\ominus = 1.67\,V$$
$$MnO_4^- \longrightarrow MnO_2 \text{ (1 M alkali)} \quad E^\ominus = 1.23\,V$$

In general, oxy-anions of transition metal are best reduced in acid solution, since this, as shown above, involves protonation. Conversely, their ions are more easily oxidised in alkaline solution. Examples appear in analysis: acidic solutions of $Co(H_2O)_6{}^{2+}$ and $Mn(H_2O)_6{}^{2+}$ are stable in air, but the precipitates of both hydroxides may be recognised since they darken rapidly on exposure to air by oxidation to the +III oxidation state. This principle is also utilised in the preparation of compounds of transition elements in high oxidation states such as MnO_4^-, CrO_4^{2-}, which are usually carried out in alkaline conditions.

Ligand substitution also alters redox potentials. Hydrated iron(III) has a potential of 0.77 V, and will oxidise iodide to iodine ($E^\ominus = 0.5\,V$). However, if the cyano-complex is made, the +III oxidation state is stabilised and the potential drops to 0.36 V. Hence hexacyanoferrate(III) will not oxidise iodide. More spectacularly, the redox potential $Co(H_2O)_6{}^{3+} + e^- \longrightarrow Co(H_2O)_6{}^{2+}$ $E^\ominus = 1.82\,V$ is high enough to allow cobalt(III) to oxidise water, and indeed solutions of cobalt(III) fluoride do release oxygen. If the ammino-complex is made, the potential drops markedly; $Co(NH_3)_6^{3+} + e^- \longrightarrow Co(NH_3)_6^{2+}$ $E^\ominus = 0.1\,V$. Hence, ammino-complexes are perfectly stable in water.

11.4 SUMMARY OF CHEMISTRY OF INDIVIDUAL ELEMENTS

11.4.1 Scandium

In its compounds, scandium exhibits a +III oxidation state only, because it loses $4s^2$ and $3d^1$ electrons to form an 'argon core' electron structure (like K^+, Ca^{2+}). In its chemical properties it is not unlike a group III element between aluminium and gallium (see chapter 4).

11.4.2 Titanium

This is an abundant element, occurring extensively as rutile, TiO_2. It is extracted for use in alloy steels and for use as the pure metal in aircraft and nuclear reactors, since it has a low density and is very resistant to corrosion at ordinary temperatures.

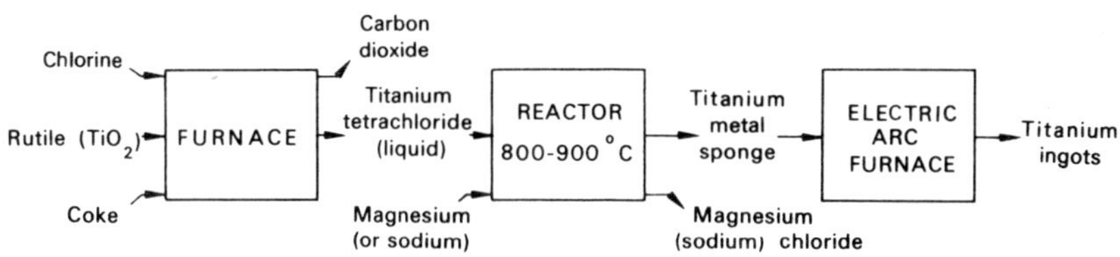

Fig. 11.6 Extraction of titanium.

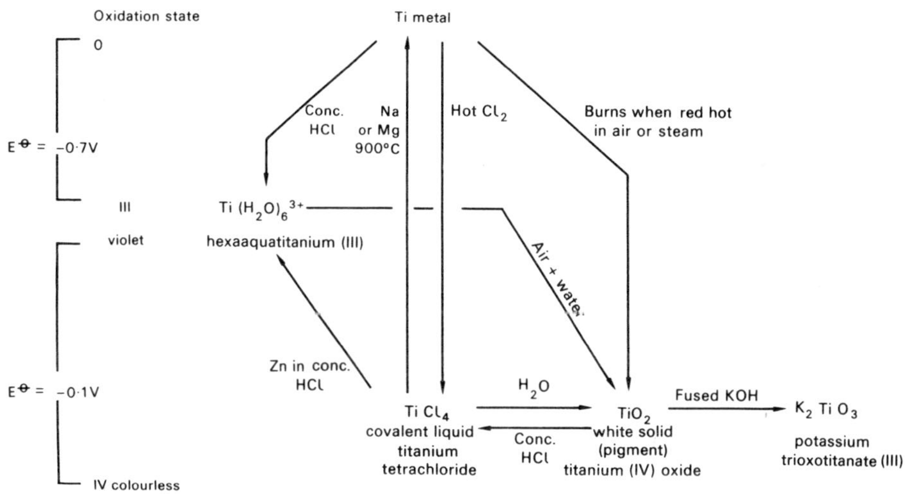

Fig. 11.7 Chemistry of titanium.

11.4.3 **Vanadium**

Vanadium is used principally in alloys with steel, especially those requiring toughness at high temperatures (high-speed cutting tools). It is a typical metal, unreactive at room temperature. The maximum oxidation state corresponds to loss of all 4s and 3d electrons, but in its higher oxidation states the ion is always strongly co-ordinated to oxygen so that V^{5+} and V^{4+} are never obtained. The colours of the various oxidation states are nicely shown by reducing a solution of ammonium vanadate(V) in acid with zinc. All the oxidation states are obtained successively, and each colour can be seen.

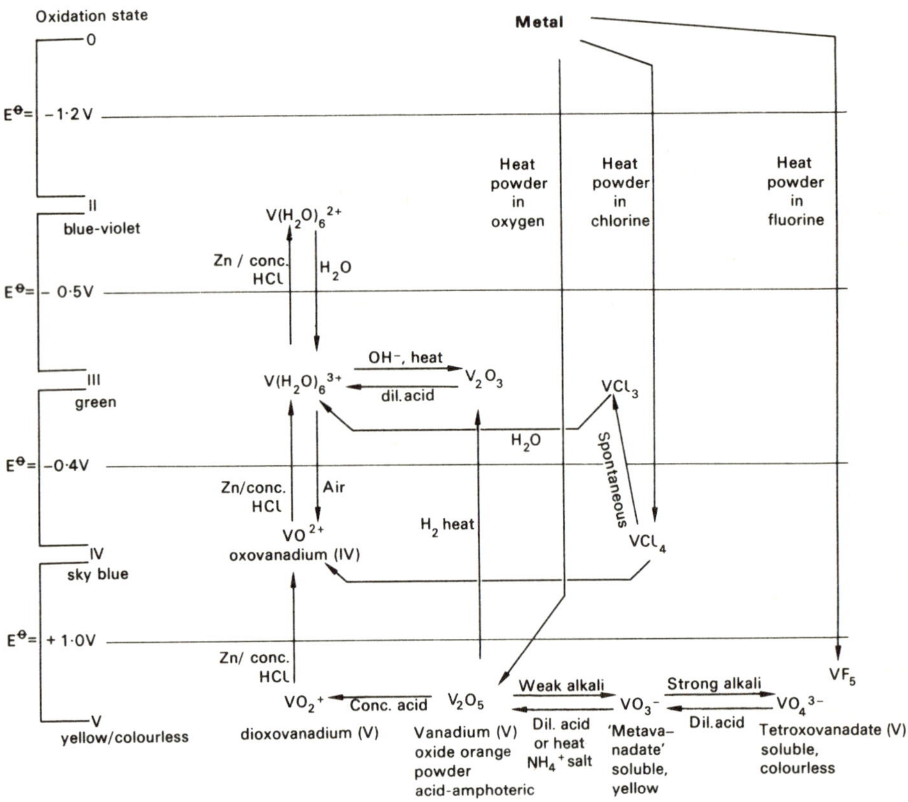

Fig. 11.8 Chemistry of vanadium.

11.4.4 **Chromium**

Chromium is a well-known, silvery metal, which takes a high polish. It is used in plating and in alloys where hardness is important.

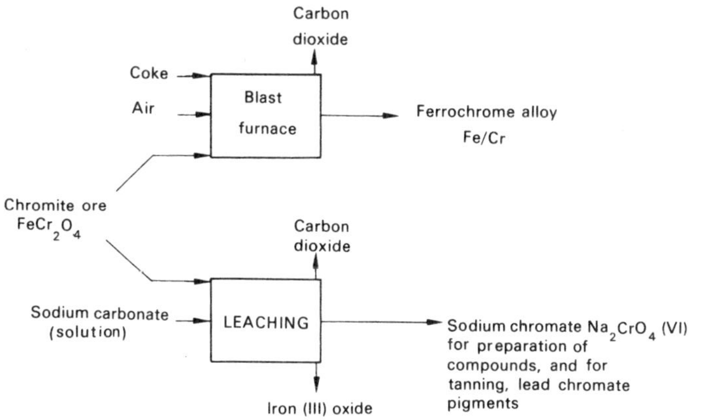

Fig. 11.9 Extraction of chromium.

Compounds of chromium are strongly coloured, and this gives the element its name. The names 'chromate'(VI) and 'dichromate'(VI) are used to refer to CrO_4^{2-} and $Cr_2O_7^{2-}$ respectively, although these are not fully systematic. The

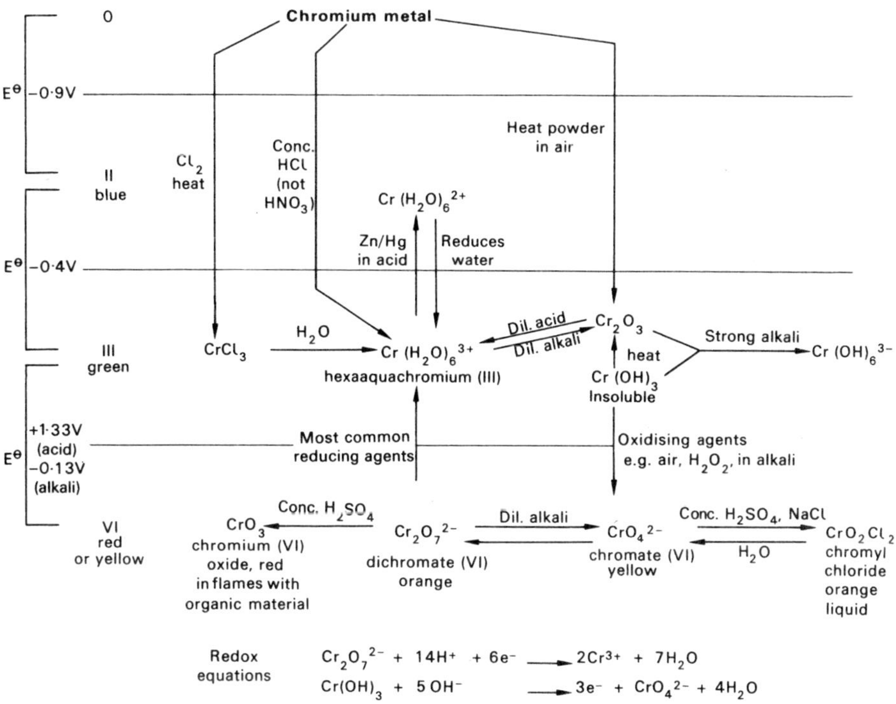

Fig. 11.10 Chemistry of chromium.

interconversion is an association/dissociation, *not* an oxidation/reduction, since the oxidation state does not change.

$$2CrO_4^{2-} + 2H^+ \longrightarrow Cr_2O_7^{2-} + H_2O$$

In chromium, the lower oxidation state of +III is now the more stable, unlike the previous elements, titanium and vanadium, in which it is the higher oxidation state which possesses the greatest stability.

11.4.5 **Manganese**

This metal, like chromium, is used to increase the toughness of steel alloys. Its chief ore is pyrolusite, MnO_2, which is itself used in dry cells as a depolariser.

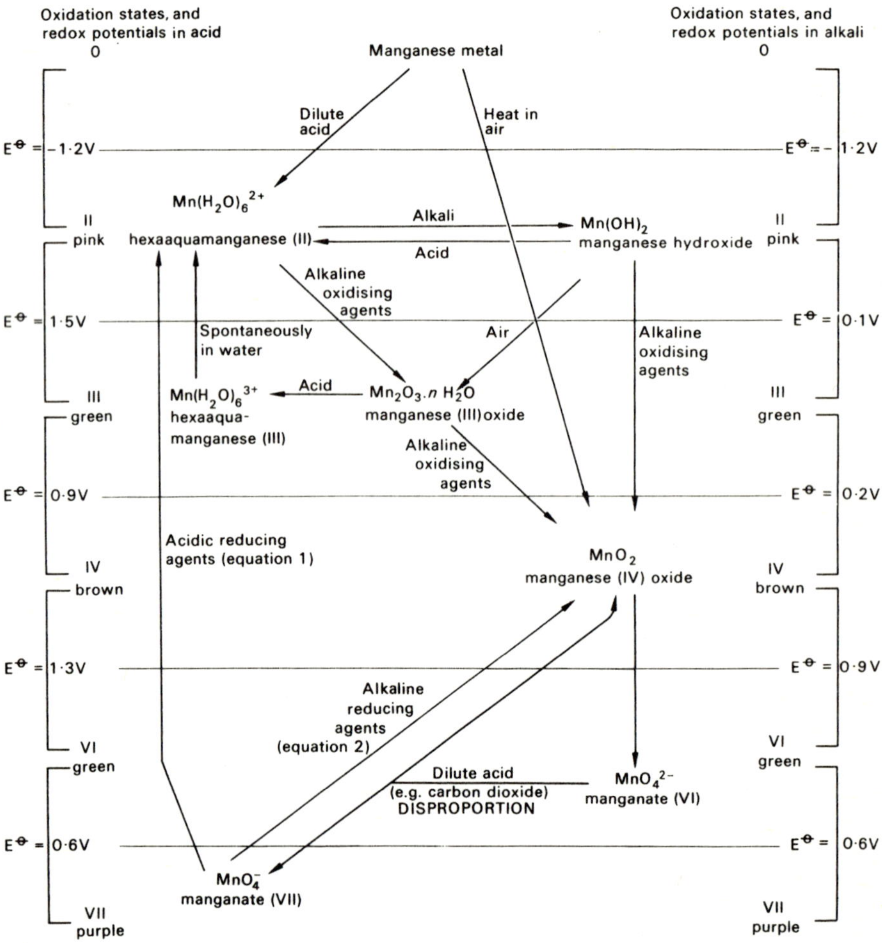

Fig. 11.11 Chemistry of manganese.

Normally the metal is not isolated but the mixed oxide ores of iron and manganese are reduced together, using carbon.

$$Fe_2O_3/MnO_2 + C \longrightarrow Fe/Mn + CO$$

Potassium manganate(VII) ($KMnO_4$, strictly tetraoxomanganate(VII)) is the compound of highest oxidation state known in the first transition series. It is a powerful oxidising agent with an intense purple colour. Since it is reduced in acid solution to almost colourless $Mn(H_2O)_6{}^{2+}$ ions, it is particularly suitable as an oxidising agent in titration.

$$MnO_4{}^- + 8H^+ + 5e^- \longrightarrow Mn^{2+} + 4H_2O$$

The end-point is clearly shown by the appearance of a pink colour, if the permanganate is in the burette. Iron(II), oxalate, iodide, and nitrate(III) are amongst the substances which can be titrated against permanganate. The disadvantage of the reagent is that aqueous solutions are not stable over long periods, and re-standardisation is continually necessary for accurate work.

$$KMnO_4 + H_2O \longrightarrow KOH + MnO_2 + O_2$$

11.4.6 **Iron**

Iron is the second most abundant metal in the earth's crust, occurring principally as its oxides Fe_2O_3 (haematite) and Fe_3O_4 (magnetite). In the U.K. it also commonly occurs as $FeCO_3$ (siderite). Since iron is more easily extracted than aluminium, the commonest metal, it is in very wide use. Pure iron is very soft and therefore more often used as an alloy, particularly as steel, in which carbon is present as cementite, Fe_3C.

Because of its very wide importance, many processes and variations have been developed for the production of iron and steel, the principal ones being shown in fig. 11.12.

The chief disadvantage of iron is that it rusts rapidly, more rapidly than other metals of a similar redox potential. Figure 11.13 explains why. A short-circuited shell is set up, with electrons being transferred from the iron to the water. The product first formed, $Fe(OH)_2$, cannot adhere and protect the surface since further oxidation forces more hydroxide into its lattice forming first iron (III) hydroxide, $Fe(OH)_3$, and then hydrated iron(III) oxide ($Fe_2O_3.nH_2O$). Protection can be achieved by a coating of a more reactive metal (e.g. zinc) which dissolves preferentially, or else by coatings such as tin (cans of food) or paint, which must remain unscratched to be effective.

The best specific test for iron salts is the formation of Prussian blue, a deep-blue insoluble substance. This is formed when either aqueous solutions of iron(II) salts are added to hexacyanoferrate(III) *or* when solutions of iron(III) salts are added to hexacyanoferrate(II). Iron(II) oxide is unimportant, being spontaneously and rapidly oxidised in air to tri-irontetroxide, Fe_3O_4.

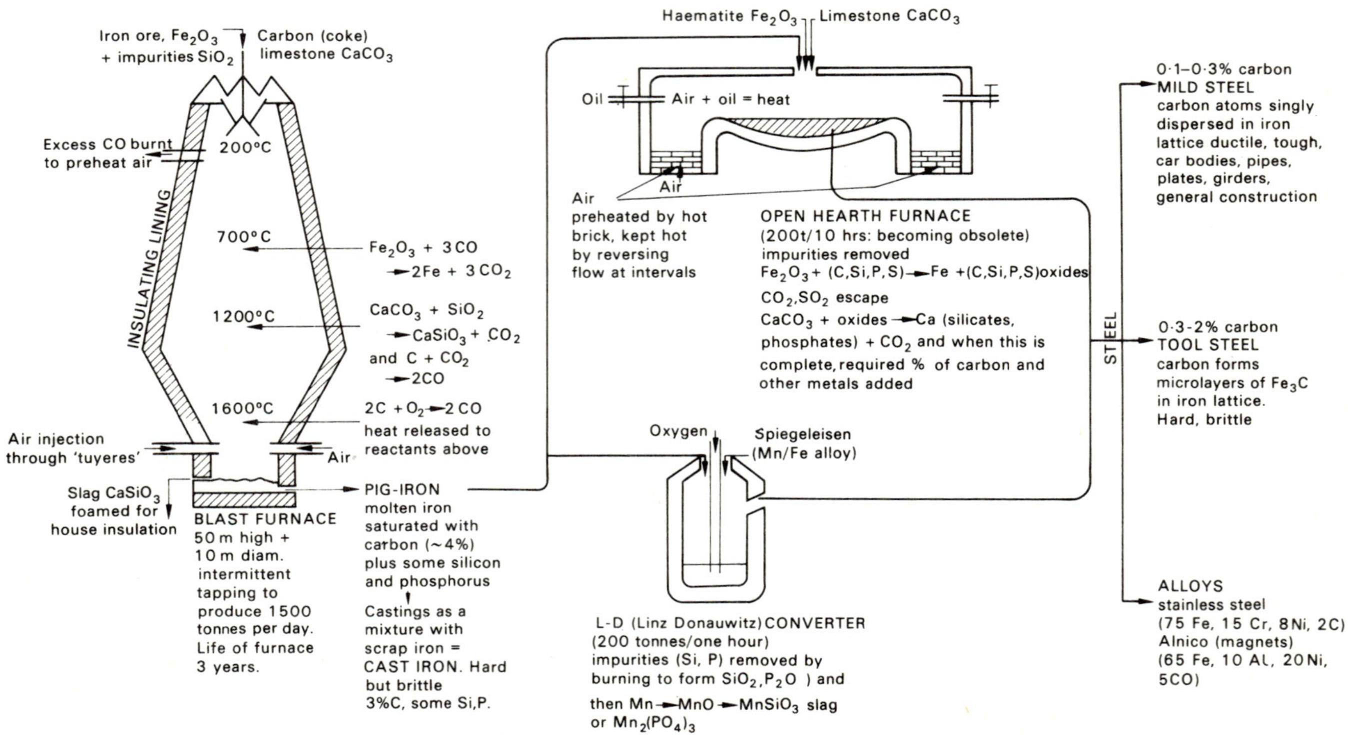

Fig. 11.12 Iron and steel.

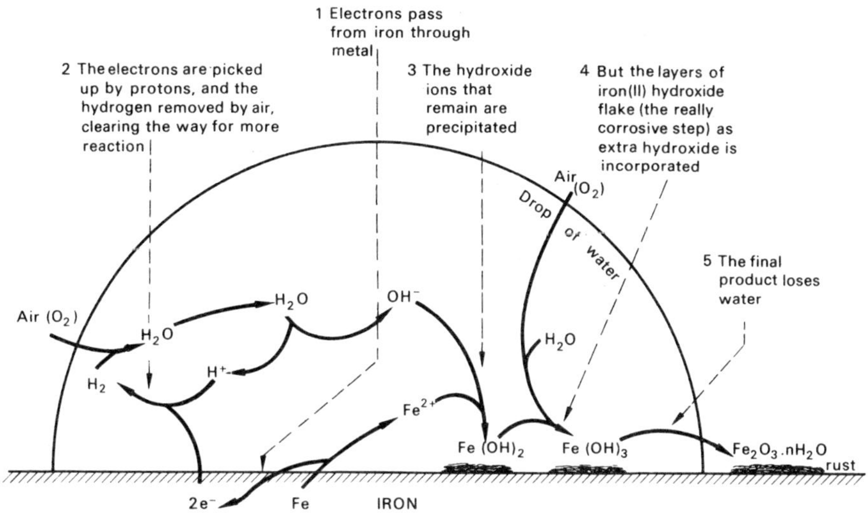

Fig. 11.13 The rusting of iron (schematic).

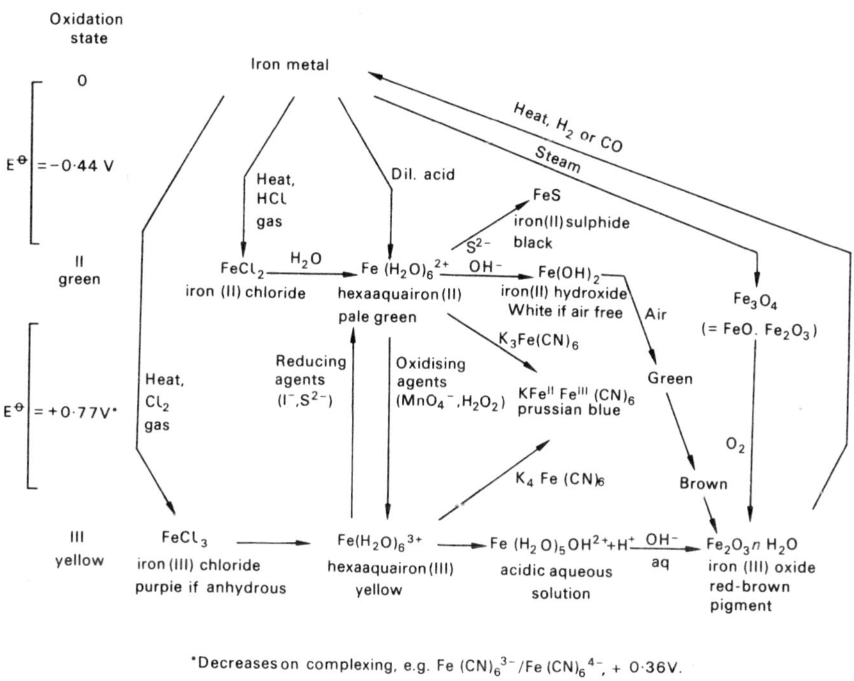

Fig. 11.14 Chemistry of iron.

11.4.7 **Cobalt**

Cobalt is used in a few alloys such as alnico, several of which are designed for use at high temperatures. It occurs in low-grade ores as sulphide or arsenide from which its extraction is complicated. Its compounds, particularly its oxide, are used in pottery glazes for producing blues, and they are also important as catalysts for organic oxidations in industry.

The chemistry of its compounds is chiefly interesting for the changes in colour and redox potential arising from change of ligand (see section 11.3). The former is put to use in self-indicating drying agents. A little cobalt chloride is added to the actual drying agent, silica gel. The following equilibrium is set up:

$$(\textit{damp})\ \underset{\text{pink}}{2Co(H_2O)_6^{2+}}\ 2Cl^- \rightleftharpoons \underset{\text{intense blue}}{Co(H_2O)_6^{2+}\ (CoCl_4)^{2-}}\ (\textit{dry}) + 6H_2O\uparrow$$

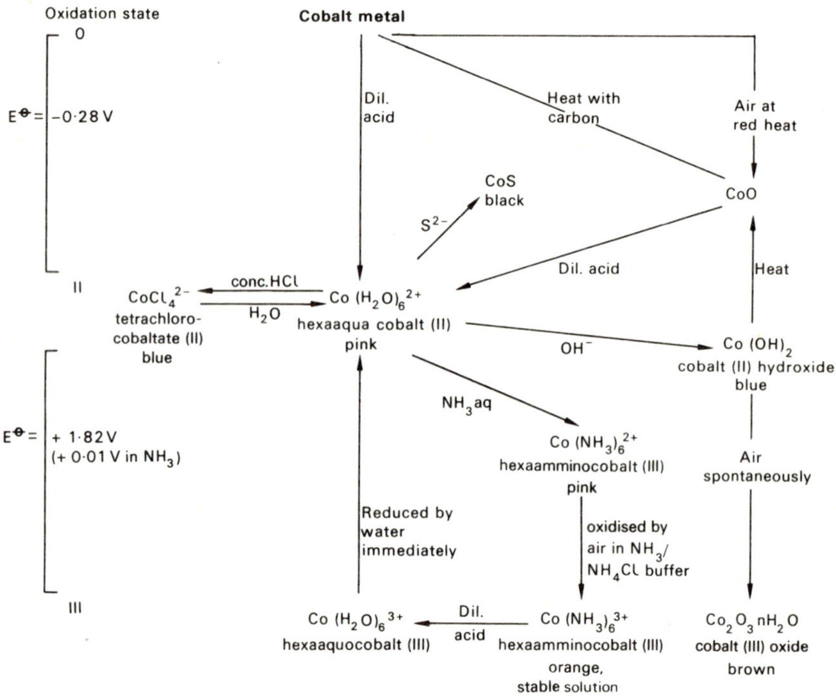

Fig. 11.15 Chemistry of cobalt.

11.4.8 **Nickel**

Nickel occurs as its sulphide in association with other metal sulphides. Several purification stages involving flotation and crystallisation yield material which is 99 per cent pure sulphide. The ore is then reduced. The principal uses of nickel include stainless steel (with Fe), magnetic alloys (with Co), and coinage alloys (with Cu).

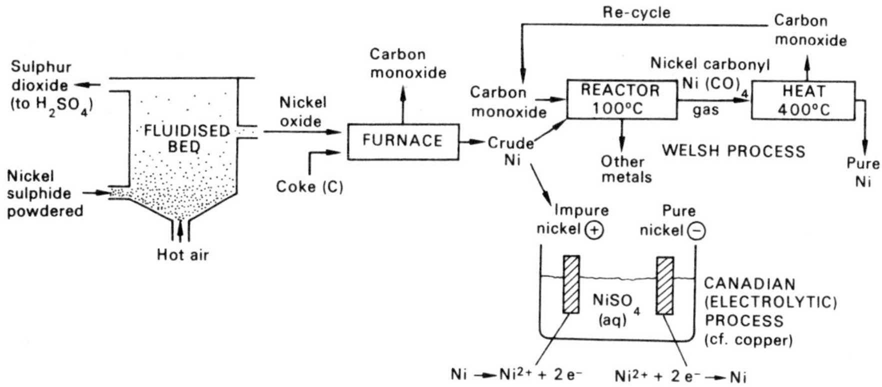

Fig. 11.16 Extraction of nickel.

The chemistry of nickel is almost exclusively that of nickel(II). Higher oxides are known but are unstable. Nickel, like cobalt, iron and chromium, is not attacked by concentrated acids, owing to the formation of a passive oxide layer.

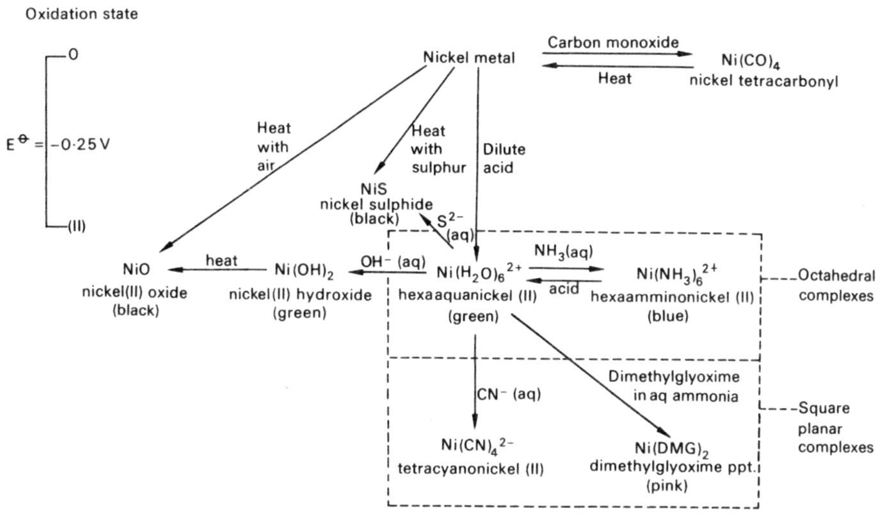

Fig. 11.17 Chemistry of nickel.

11.4.9 **Copper**

Copper occurs principally as copper pyrites, $CuFeS_2$, and copper glance, Cu_2S. It is extracted as shown in Fig. 11.18 and refined electrolytically. High-purity copper, as produced by this process, is essential for its use as an electrical or thermal conductor, since small amounts of impurities lower the conductivity dramatically. In the electrolytic purification, impure copper is used as the anode in the electrolysis of copper sulphate. The copper dissolves ($Cu \longrightarrow Cu^{2+} + 2e^-$) as well as any reactive metal impurities (Zn, Fe), while less reactive metals (Ag, Au) fall to the bottom as anode mud. The cathode is pure copper, since copper ions discharge preferentially leaving zinc ions, etc. in solution. Besides its use in conductors, copper is used for pipes and weather-resistance in buildings, and in various alloys such as cupro-nickel (coinage), brass (with zinc) and bronze (with tin).

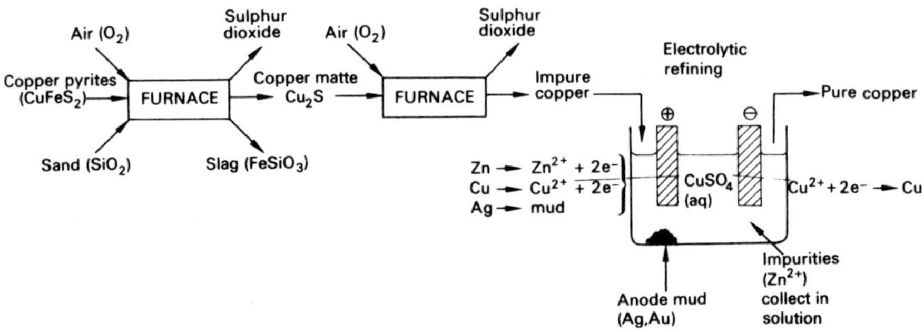

Fig. 11.18 Extraction of copper.

Although the +II oxidation state of copper is the stable one under atmospheric conditions, the +I state is also important. The Cu^+ ion is not stable as an aqueous complex, and disproportionates, a result predicted by the following redox potentials.

$$Cu^+(aq) + e^- \longrightarrow Cu \qquad E^\ominus = +0.52\,V$$
$$Cu^{2+}(aq) + e^- \longrightarrow Cu^+(aq) \qquad E^\ominus = +0.16\,V$$
hence
$$2Cu^+(aq) \longrightarrow Cu + Cu^{2+}(aq) \qquad E^\ominus = +0.36\,V$$

To be stable in the presence of water, therefore, copper(I) must either be complexed (as in $CuCl_2^-$) or form part of an insoluble compound (as in CuI, Cu_2O or Cu_2S).

Copper(II) ions exist in water as the familiar blue aquo-complex, $Cu(H_2O)_6^{2+}$. This complex has the shape of a distorted octahedron (fig. 11.19), four of the water ligands forming a square, and the remaining two at slightly greater distances. This distortion is preserved in the ammino-complex, $Cu(NH_3)_4(H_2O)_2^{2+}$, in which the ammonia ligands form the square; and in crystalline copper(II) sulphate, $Cu(H_2O)_4^{2+}\ SO_4^{2-}.H_2O$, in which the sulphate ions are the greater distance away.

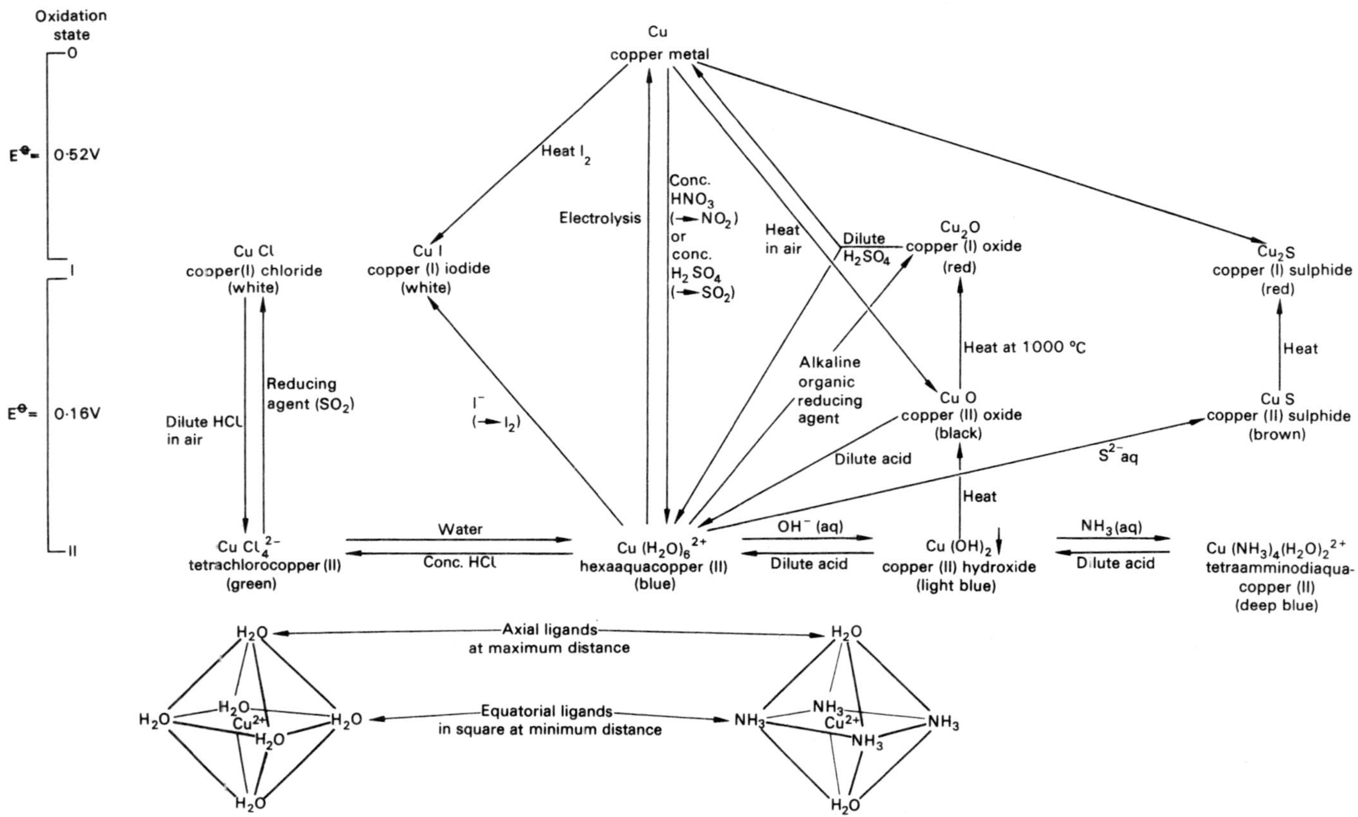

Fig. 11.19 Chemistry of copper.

11.4.10 **Zinc**

Zinc blende, ZnS, is the principal ore, although zinc carbonate is also found. Since zinc is more volatile than any likely contaminants (Cu, Pb) it is obtained relatively pure from the furnace when it distils out. Further purification can be achieved by vacuum distillation. Its principal uses are for the protection of steel by galvanising, for casting alloys including brass, and in dry cells.

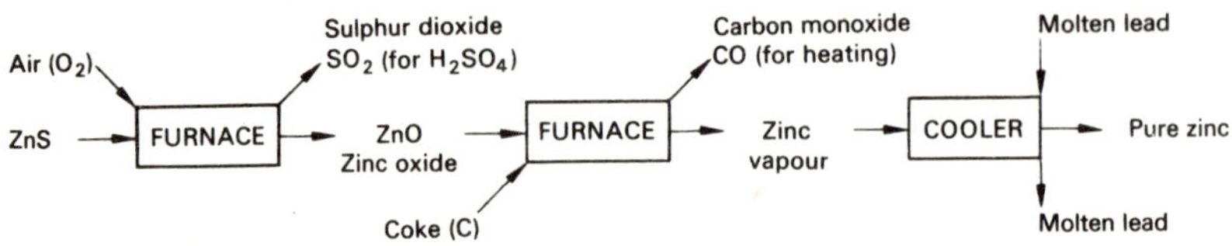

Fig. 11.20 Extraction of zinc.

Zinc exhibits only one oxidation state in its compounds, owing to its electronic structure of $3d^{10}\ 4s^2$. Loss of the 4s electrons leaves the stable arrangement of a full outermost shell. This arrangement also gives it the smallest ionic radius of any element in the series, so that its oxide and hydroxide are definitely amphoteric; 4 co-ordinate compounds are common partly because of the crowding arising from the presence of six ligands, and partly because four ligands (donating two electrons each) will exactly fill the available 4s and 4p orbitals.

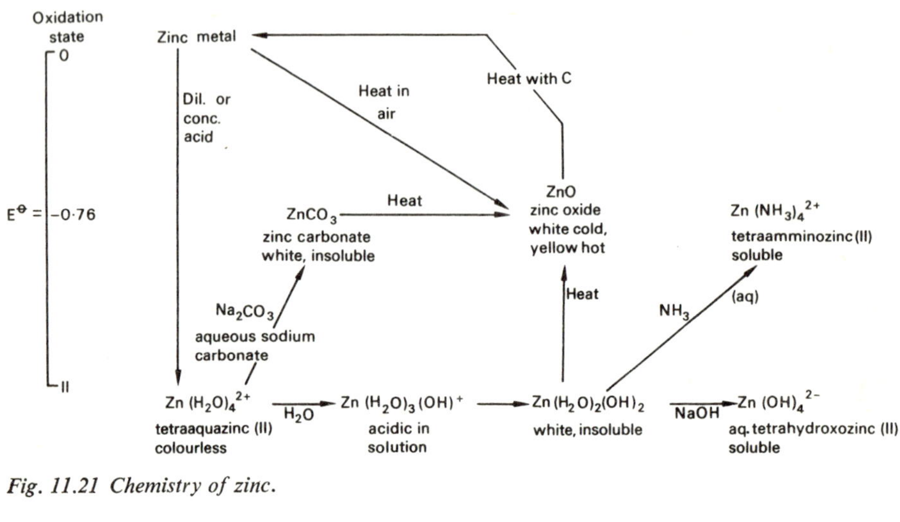

Fig. 11.21 Chemistry of zinc.

QUESTIONS

1 Choose four directly observable properties of transition metals or their compounds, and show how these properties are derived from the characteristic electronic structure of this group.

2 Show how the following pairs of properties are connected. The first property may depend on the second, or vice versa, or both may have a common cause.

(a) The number of electrons readily removed from a transition metal atom is equal to the number in the 3d and 4s orbitals.	The maximum oxidation state of the elements in the series Sc–Ti–V–Cr–Mn is 3–4–5–6–7.
(b) Coloured solutions usually contain transition metal ions.	Complex ion formation is characteristic of transition metals.
(c) Transition metals show variable oxidation states.	Industrial catalysts are often chosen from elements Ti $\longrightarrow$ Cu.
(d) The redox potential $Co^{3+}/Co^{2+}(aq)$ is sufficient to oxidise water.	$Co(NH_3)_6{}^{3+}$ is stable in alkaline, but not in acid solutions.
(e) There are five isomers of the formula $PtCl_4.6NH_3$.	A fixed number of ligands coordinate to a central atom.

3 Write balanced equations for *three* examples of each of the following, in ionic form or otherwise as appropriate:
(a) a half equation showing a transition metal reduced by three oxidation states or more;
(b) a ligand replacement reaction resulting in a colour change;
(c) a ligand replacement reaction resulting in a large change of redox potential;
(d) an industrial process catalysed by a transition metal;
(e) a transition metal dissolving in dilute hydrochloric acid;
and *two* examples of each of the following:
(f) a disproportionation involving a transition metal;
(g) the preparation of a compound with a transition metal in an oxidation state of five or more;
(h) the precipitation of a coloured hydroxide;
(i) the reduction of water by an aqueous transition metal ion;
(j) the oxidation of iodide to iodine by a transition metal ion.

4 Describe the extraction of iron from haematite. Explain as if to a person with a smattering of scientific knowledge
(a) why it is cheaper to fabricate complex articles (such as cars) out of iron rather than aluminium;
(b) why steel is tougher than pure iron;
(c) why galvanising prevents the corrosion of iron;
(d) why painting over rust does not prevent further corrosion beneath.

5 Identify the characteristics of the following that make them suitable for the specified purpose. How does this characteristic arise from the more fundamental properties of electronic structure, redox potential, etc.
(a) Manganate(VII) is useful for redox titrations.

(b) Gold is traditionally the metal of value.
(c) Chromate(VI) is used in breathalysers.
(d) Titanium is used for toughness where costs are not important.
(e) Cobalt(II) chloride is added to silica gel drying agent as a moisture indicator.

6 The following statements are not correct in all respects. There may be exceptions, ambiguities, omissions or deliberate mistakes. Write out a correct version of the statement and comment on the theoretical background to it.
(a) The maximum oxidation state of transition metals is equal to the number of d electrons.
(b) Copper(I) salts are not stable in the presence of water.
(c) The +(II) oxidation state of transition metals is the one stable in aqueous solution.
(d) Transition metals, being in the centre of the periodic table, have amphoteric oxides.
(e) The catalytic activity of transition metals is an instance of their ability to be reversibly oxidised.
(f) Transition metal complexes have an octahedral shape.
(g) Chromium compounds are green.
(h) The impurities in copper refining collect as anode mud.
(i) The hydroxides $M(OH)_3$ redissolve in excess sodium hydroxide.
(j) Transition metal ions are coloured.

12
Aqueous Ionic Equilibria

Simple tests for ions in solution

12.1 INTRODUCTION

Many of the substances dealt with in this book are ionic compounds formed when metals react with non-metals. The cations and anions respectively formed by these are easily recognised by a small number of test-tube reactions. Although these tests are mentioned individually under the elements concerned, they are gathered here for comparison. These particular tests are well established because they satisfy three conditions:

(1) they are simple and quick to perform;
(2) they give an easily observable result, by colour change, or precipitate, or odour;
(3) the one test yields different results for different ions.

Modern industry does not use these tests, since its analyses involve complex mixtures which can only be separated with specialised skills and sophisticated equipment. For example, some £15000 might be spent on a spectrophotometer, in which the light emitted by a vaporised sample is analysed automatically for the frequencies emitted by each element. The laboratory flame test is a crude version of this; sodium compounds introduced into a bunsen flame produce a brilliant yellow colour; calcium, red; barium, green; potassium, lilac; and so on.

In the school laboratory, relatively simple compounds can be analysed by carrying out a number of solution reactions. In making a solution of an unknown substance, the following are tried (in order of preference): water, dilute hydrochloric acid, concentrated hydrochloric acid, dilute nitric acid and finally concentrated nitric acid. The further along the order of preference, the more likely is the solvent to interfere with later tests. For instance, nitric acid will oxidise hydrogen sulphide to a white precipitate of sulphur, so masking any sulphide precipitate (see section 12.2).

12.2 PRECIPITATION REACTIONS OF SULPHIDES

The solubility of a substance can be described by its solubility product. When copper(II) sulphide is precipitated, for instance, the solid precipitate is in contact with a saturated solution of copper(II) sulphide, albeit a very weak one.

The equilibrium can be summarised:

$$Cu^{2+}(aq) + S^{2-}(aq) \rightleftharpoons CuS(s) \quad (\rightleftharpoons \text{ denotes that equilibrium is displaced well to the right})$$

It is found that the product of the molar concentrations of Cu^{2+} and S^{2-} is 10^{-36} $(mol\ dm^{-3})^2$ and this is known as the solubility product. A sparingly soluble substance, such as $Ca(OH)_2$, might have a value of 10^{-5}, whereas a really insoluble one, such as HgS, might have as low a value as 10^{-52}. This last value means that $[Hg^{2+}] = [S^{2-}] = 10^{-26}$ mol dm^{-3}, about one ion per hundred litres!

Similarly, the value for copper(II) sulphide implies that:

a 1 M solution of Cu^{2+} will produce a precipitate as soon as the concentration of S^{2-} exceeds 10^{-36} M
a 10^{-1} M solution of Cu^{2+} will produce a precipitate as soon as the concentration of S^{2-} exceeds 10^{-35} M
a 10^{-2} M solution of Cu^{2+} will produce a precipitate as soon as the concentration of S^{2-} exceeds 10^{-34} M
and so on.

Sulphides precipitated by H_2S *in acid solution*			*Sulphides precipitated by* H_2S *in neutral solution*			*Sulphides not precipitated by* H_2S		
Formula	*Solubility product*	*Colour*	*Formula*	*Solubility product*	*Colour*	*Formula*	*Solubility product*	*Colour*
Ag_2S	10^{-50}	brown	CoS	10^{-21}	black	CaS	10^{-5}	white
Bi_2S_3	10^{-97}	brown	FeS	10^{-18}	black			
CdS	10^{-28}	yellow	MnS	10^{-10}	pink	Na_2S	10^{-2}	white
CuS	10^{-36}	brown	NiS	10^{-19}	black			
HgS	10^{-52}	red	ZnS	10^{-22}	white	Fe^{3+} does not precipitate a sulphide but is reduced to Fe^{2+}, releasing sulphur (pale yellow)		
PbS	10^{-28}	black						
SnS	10^{-25}	brown						

Table 12.1 Sulphide precipitates.

Hydrogen sulphide is a weak acid; nevertheless, at pH 7 it produces an ample concentration of sulphide ions (10^{-7}M) to precipitate copper(II) sulphide. However, the test is carried out in two stages. Firstly, the test solution is acidified with dilute hydrochloric acid, and hydrogen sulphide is then passed through. This has the effect of lowering the concentration of sulphide ions still further:

$$2H^+(aq) + S^{2-}(aq) \rightleftharpoons H_2S(aq)$$

By Le Chatelier's principle, an increase in $[H^+]$—left-hand side—will cause a shift to the right. The sulphide concentration is actually reduced to about 10^{-22}M. This is sufficient to precipitate the sulphides listed in the first column of table 12.1 but none of the others.

In the second stage, the solution is just neutralised with ammonia. This raises the sulphide concentration to 10^{-7}M by removing hydrogen ions, replaced by dissociation of the hydrogen sulphide. The sulphides in the second column are then precipitated.

12.3 PRECIPITATION REACTIONS OF HYDROXIDES

Metal ions which are strongly hydrated in solution act as weak acids. If a solution containing hydroxide ions is added to a solution of such metal ions, protons are removed.

$$\underset{\text{hexaaquametal(II)}}{M(H_2O)_6^{2+}} \xrightarrow{OH^-} M(H_2O)_5(OH)^+ \xrightarrow{OH^-} \underset{\text{tetraaquadihydroxometal(II)}}{M(H_2O)_4(OH)_2}$$

The last complex in the sequence, being uncharged, interacts only weakly with the dipole of the water molecule, and is therefore usually precipitated. The only cations not precipitated in this way are the group I metals, sodium to caesium.

This neutral complex may react further, depending on the source of the hydroxide ion. If sodium hydroxide is used, the concentration of hydroxide ions is high, and addition of excess reagent may dissolve the complex, as it now becomes negatively charged.

$$M(H_2O)_4(OH)_2 \xrightarrow{OH^-} M(H_2O)_3(OH)_3^- \dashrightarrow \underset{\text{hexahydroxometallate(II)}}{M(OH)_6^{4-}}$$

This is particularly likely to happen where the O–H bond is polarised by a metal ion of small size and high charge. For a full discussion see chapter 4.

On the other hand, if aqueous ammonia is used, the concentration of hydroxide ions is low (about 10^{-5} M in 1 M ammonia)

$$NH_4^+ + OH^- \rightleftharpoons NH_3 + H_2O$$

It is then possible for the neutral complex to exchange ammonia for water or hydroxide, and to redissolve in excess alkali.

$$Cu(H_2O)_6^{2+} \xrightarrow{OH^-} Cu(H_2O)_4(OH)_2 \xrightarrow{NH_3} Cu(NH_3)_4(H_2O)_2^{2+}$$

The responses of different cations to aqueous sodium hydroxide or aqueous ammonia can be divided into various groups which are related to the position of the elements in the periodic table (see fig. 12.1).

Group A: These elements form very soluble hydroxides which are not precipitated by either reagent.

Group B: These elements form white, sparingly soluble hydroxides, which are precipitated by adding either sodium hydroxide or ammonia solution to a solution of the cation. The precipitates are often faint and slow to form.

Group C: These elements form insoluble hydroxides, which are precipitated by adding either reagent to a dilute solution of the cation. $Mg(OH)_2$, $Sc(OH)_3$, $Ti(OH)_4$ (white); $V(OH)_3$ (green).

Group D: These elements form insoluble hydroxides, which are precipitated by adding either reagent to a dilute solution of the cation. Hydroxides of the lower oxidation state are, however, converted to the higher oxidation state on standing in air, $Cr(OH)_2$ (yellow-brown) becoming green $Cr(OH)_3$ (see group F); $Mn(OH)_2$ (pink) becoming brown Mn_2O_3, hydrated; $Fe(OH)_2$ (pale green) becoming darker and eventually brown $Fe(OH)_3$; $Co(OH)_2$ (blue) becoming grey Co_2O_3, hydrated.

Group E: These elements form insoluble hydroxides, which are precipitated by either reagent, but which redissolve in excess ammonia solution to form the ammino complex. This group do not redissolve in sodium hydroxide, except for zinc.

$Ni(OH)_2$ (green) becoming blue $Ni(NH_3)_6{}^{2+}$.
$Cu(OH)_2$ (pale blue) becoming dark blue $Cu(NH_3)_4(H_2O)_2{}^{2+}$.
'AgOH', properly $Ag_2O.\,nH_2O$ (brown), becoming colourless $Ag(NH_3)_2{}^{+}$.
$Zn(OH)_2$ (white) becoming colourless $Zn(NH_3)_4{}^{2+}$.
$Cd(OH)_2$ (white) becoming colourless $Cd(NH_3)_4{}^{2+}$.

Group F: These elements form white insoluble hydroxides (except $Cr(OH)_3$, grey-green) which are precipitated by either reagent, but will redissolve in excess sodium hydroxide to form colourless solutions. This group do not redissolve in ammonia except for zinc.

$X(OH)_2 \longrightarrow X(OH)_4{}^{2-}$ Be, Zn
$X(OH)_3 \longrightarrow X(OH)_6{}^{3-}$ Al, Ga, Cr
$X(OH)_2 \longrightarrow X(OH)_6{}^{4-}$ Ge, Sn, Pb

Group G: The 'hydroxides' of these elements are unstable and are only known in their dehydrated forms as the familiar mineral acids. For example

'$C(OH)_4$' $\equiv H_2CO_3 + H_2O$
'$N(OH)_5$' $\equiv HNO_3 + 2H_2O$
'$S(OH)_6$' $\equiv H_2SO_4 + 2H_2O$

Ammonium ions form a group of their own in that the gas ammonia is given off, which may be detected by its characteristic smell.

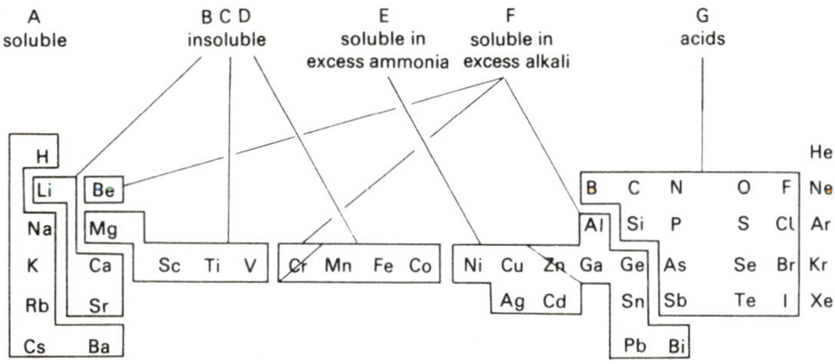

Fig. 12.1 Solubility of hydroxides and the periodic table.

12.4 OTHER PRECIPITATION REACTIONS OF CATIONS

Although carbonates are usually insoluble (except those of Na^+, K^+, NH_4^+), the precipitation of carbonates is not often used as a test and although carbonate precipitates are coloured, unlike hydroxides they do not dissolve in an excess of the reagent. Hence less information is available from the addition of sodium carbonate than from addition of sodium hydroxide, and the reaction is not used as a test. The elements barium, calcium, lead and silver do form several insoluble salts with other anions, however. A solution of the appropriate sodium salt can be added to the test solution (see table 12.2).

	Sodium salt used			
Cation	*Chloride*	*Chromate(VI)*	*Iodide*	*Sulphate(VI)*
Barium	no ppt.	yellow ppt. insoluble in acetic acid	no ppt.	white ppt.
Calcium	no ppt.	yellow ppt. soluble in acetic acid	no ppt.	faint white ppt.
Lead	white ppt. soluble in hot water	yellow ppt.	yellow ppt.	white ppt.
Silver	white ppt.	red ppt.	light yellow ppt.	no ppt.

Table 12.2 Insoluble compounds of Ba^{2+}, Ca^{2+}, Pb^{2+} *and* Ag^+.

12.5 REACTIONS OF ACIDS WITH ANIONS

Anions are capable in all cases of acting as bases, that is, they will accept protons. In so doing the parent acid is formed, and, if this is gaseous, there is often visible effervescence or detectable odour, particularly on warming.

$$X^-(aq) + H^+(aq) \rightleftharpoons HX(aq) \longrightarrow HX(g)$$

If dilute hydrochloric acid is used, it will be a sufficiently strong acid to protonate most anions. In the test the acid is added to the solid sample and one of the following gases may be observed.

Sample	*Reaction*	*Observation*
Carbonate	$CO_3^{2-} + 2H^+ \longrightarrow H_2O + CO_2$	Vigorous effervescence. Gas turns lime water milky
Chlorate(I)	$OCl^- + 2H^+ + Cl^- \longrightarrow H_2O + Cl_2$	Green gas, turns starch/potassium iodide paper purple-black
Nitrate(III) (Nitrite)	$NO_2^- + H^+ \longrightarrow HNO_2$ $HNO_2 \longrightarrow N_2O_3 \longrightarrow NO + NO_2$	Green nitric(III) acid, first formed, decomposes rapidly to blue dinitrogen trioxide (solution) and brown nitrogen dioxide (evolved)
Sulphate(IV) (Sulphite)	$SO_3^{2-} + 2H^+ \longrightarrow H_2O + SO_2$	Strong pungent smell, choking. Gas turns acidified dichromate paper from orange to green
Sulphide	$S^{2-} + 2H^+ \longrightarrow H_2S$	Bad egg smell. Gas turns moist lead acetate paper black

Table 12.3 Gases evolved during testing with dilute acid.

Other protonation reactions may occur without evolution of gas, and these are often best seen if solutions are tested.

$$\underset{\text{chromate(VI) yellow solution}}{CrO_4^{2-}} + 2H^+ \longrightarrow H_2CrO_4 \longrightarrow \tfrac{1}{2}\{\underset{\text{dichromate(VI) orange solution}}{H_2Cr_2O_7} + H_2O\}$$

$$\underset{\text{benzoate or salts of other acids of high molecular mass colourless solution}}{C_6H_5CO_2^-} + H^+ \longrightarrow \underset{\text{benzoic acid white ppt.}}{C_6H_5CO_2H}$$

If concentrated sulphuric acid is used as a protonating agent, anions which correspond to strong acids will also be protonated, in addition to the above. A further reaction then occurs if the acid released is a reducing agent. In the test, the concentrated acid is added to, and warmed cautiously with, the solid sample. (*Wear safety glasses.*)

Sample	*Reaction and observation*	*Further reaction and further observation*
Fluoride*	$H^+ + F^- \longrightarrow HF$ White fumes which etch moist glassware	–
Chloride	$H^+ + Cl^- \longrightarrow HCl$ White, choking fumes	–
Bromide	$H^+ + Br^- \longrightarrow HBr$ White, choking fumes	$2HBr \longrightarrow 2H^+ + 2e^- + Br_2$ $H_2SO_4 + 2H^+ + 2e^- \longrightarrow 2H_2O + SO_2$ Some brown fumes evolved on warming
Iodine	$H^+ + I^- \longrightarrow HI$ Next stage proceeds instantly	$2HI \longrightarrow 2H^+ + 2e^- + I_2$ $H_2SO_4 + 8H^+ + 8e^- \longrightarrow 4H_2O + H_2S$ Purple iodide vapour and bad egg smell
Nitrate(V)	$H^+ + NO_3^- \longrightarrow HNO_3$ Some clear distillate	$4HNO_3 \longrightarrow 2H_2O + 4NO_2 + O_2$ Brown fumes on heating

* CARE: HYDROGEN FLUORIDE IS VERY POISONOUS—USE A FUME CUPBOARD

Table 12.4 Gases evolved during testing with concentrated acid.

12.6 PRECIPITATION REACTIONS OF SILVER IONS

Because silver forms a wide range of insoluble compounds, addition of silver nitrate solution to the test solution often produces a characteristic precipitate.

$$Ag^+ + X^- \rightleftharpoons AgX$$

Subsequent addition of dilute nitric acid allows further distinctions, because X^- will be protonated.

$$X^- + H^+ \rightleftharpoons HX$$

If HX is a very weak acid, sufficient X^- may be removed by protonation to allow the concentration of silver ions to be as high as 1 M, without exceeding the solubility product. In other words, the precipitate of silver salt will dissolve. Addition of ammonia also dissolves insoluble silver salts, by forming diamminosilver(I) ions.

$$Ag^+ + 2NH_3 \rightleftharpoons Ag(NH_3)_2^+$$

Anion	*Precipitate formed on adding* $AgNO_3$	*Effect of adding dilute* HNO_3	*Reaction with dilute* HNO_3	*Effect of adding* NH_3 *solution to precipitate first formed*
Hydroxide	brown $Ag_2O.nH_2O$	dissolves	$O^{2-} + 2H^+ \longrightarrow H_2O$	dissolves
Carbonate	cream Ag_2CO_3	dissolves	$CO_3^{2-} + 2H^+ \longrightarrow H_2O + CO_2$	dissolves
Fluoride	none	–	–	–
Chloride	white AgCl	none	–	dissolves
Bromide	cream AgBr	none	–	dissolves if conc. NH_3
Iodide	pale yellow AgI	none	–	none
Nitrate(III)	white $AgNO_2$	dissolves	$NO_2^- + H^+ \longrightarrow HNO_2$	dissolves
Chromate(VI) *or* Dichromate(VI)	red Ag_2CrO_4	dissolves	$CrO_4^{2-} + 2H^+ \longrightarrow H_2CrO_4$	dissolves
Phosphate(V)	yellow Ag_3PO_4	dissolves	$PO_4^{3-} + H^+ \longrightarrow HPO_4^{2-}$	dissolves
Sulphide	brown Ag_2S	dissolves	$S^{2-} + 2H^+ \longrightarrow H_2S$	dissolves

Sulphates(IV) and (VI) may give faint white precipitates, because the silver salts are sparingly soluble.

Table 12.5 Precipitation reaction of silver ions.

12.7 OTHER PRECIPITATION REACTIONS FOR ANIONS

Barium nitrate (or chloride) is used with nitric acid, principally as a confirmatory test for sulphate(VI). Barium sulphate(VI) is the only precipitate to remain undissolved under these conditions. Before the addition of nitric acid, a white precipitate indicates either sulphate(VI), sulphate(IV) or carbonate, but the last two dissolve in dilute acid.

12.8 REDOX TESTS FOR IONS

If a solution of potassium iodide, acidified with dilute sulphuric acid, is added to the test solution, the formation of a purple–black precipitate confirms the presence of an oxidising agent. Excess iodide may dissolve the precipitate to give a brown solution.

$$I^- \longrightarrow \tfrac{1}{2}I_2 + e^- \quad E^\ominus = -0.54\,V$$
$$I_2 + I^- \longrightarrow I_3^-$$

Any ion with a redox potential greater than 0.54 V should give a positive result.

Sample	*Redox equation*	$E^\ominus$	*Observations and comments*
Chlorate(I)	$HOCl + H^+ + e^- \longrightarrow \tfrac{1}{2}Cl_2 + H_2O$	1.64	Instantaneous
Chromate(VI) or Dichromate(VI)	$\tfrac{1}{6}(Cr_2O_7^{2-} + 14H^+ + 6e^- \longrightarrow 2Cr^{3+} + 7H_2O)$	1.33	Orange dichromate(VI) becomes green chromium(III). Breathalyser test
Copper(II)	$Cu^{2+} + I^- + e^- \longrightarrow CuI$	0.86	Blue solution loses colour, and white ppt. of copper(I) iodide formed
Iron(III)	$Fe^{3+} + e^- \longrightarrow Fe^{2+}$	0.77	A slow reaction
Manganate(VII)	$\tfrac{1}{5}(MnO_4^- + 8H^+ + 5e^- \longrightarrow Mn^{2+} + 4H_2O)$	1.52	Purple manganate(VII) decolorised
Nitrate(V)	$NO_3^- + 2H^+ + e^- \longrightarrow NO_2 + H_2O$	0.81	Very slow, except if concentrated acid is used
Nitrate(III)	$HNO_2 + H^+ + e^- \longrightarrow NO + H_2O$	0.99	Nitrogen monoxide converted to dioxide by air, so that brown fumes are seen

Table 12.6 Oxidation reactions of ions.

Sulphides are not used as a test reagent, despite having a low redox potential, because of the possibility of confusion with sulphide precipitates. Ions with reducing properties may be tested by the dropwise addition of acidified manganate(VII) which is then decolorised.

$$\tfrac{1}{5}(MnO_4^- + 8H^+ + 5e^- \longrightarrow Mn^{2+} + 4H_2O) \quad E^\ominus = 1.52\,V$$

Any ion with a redox potential of less than 1.52 V should give a positive result. (N.B. In the test the action of manganate(VII) reverses the given standard redox equations.)

Sample	*Redox equation*	$E^\ominus$	*Observations and comments*
Bromide	$\tfrac{1}{2}Br_2 + e^- \longrightarrow Br^-$	1.07	Slow reaction: may not be noticed
Chloride	$\tfrac{1}{2}Cl_2 + e^- \longrightarrow Cl^-$	1.36	Slow reaction: may not be noticed
Iron(II)	$Fe^{3+} + e^- \longrightarrow Fe^{2+}$	0.77	
Iodide	$\tfrac{1}{2}I_2 + e^- \longrightarrow I^-$	0.54	Purple precipitate or brown solution of iodine also formed
Manganese(II)	$\tfrac{1}{2}(MnO_2 + 4H^+ + 2e^- \longrightarrow Mn^{2+} + 2H_2O)$	1.23	Brown ppt. of MnO_2. The manganate(VII) is on this occasion also reduced to MnO_2
Nitrate(III)	$\tfrac{1}{2}(NO_3^- + 3H^+ + 2e^- \longrightarrow HNO_2 + H_2O)$	0.94	Brown fumes of NO_2 from side reaction $H^+ + NO_2^-$ (see table 12.3)
Sulphide	$\tfrac{1}{2}(S + 2H^+ + 2e^- \longrightarrow H_2S)$	0.15	White ppt. of sulphur
Sulphate(IV)	$\tfrac{1}{2}(SO_4^{2-} + 4H^+ + 2e^- \longrightarrow H_2SO_3 + H_2O)$	0.17	
Tin(II)	$\tfrac{1}{2}(Sn^{4+} + 2e^- \longrightarrow Sn^{2+})$	0.15	

Table 12.7 Reduction reactions of common ions.

12.9 SPECIFIC TESTS FOR IONS

Specific tests for various ions are given in table 12.8 and in the chapters describing the reactions of the elements in detail. Some of these involve precipitation:

$$\text{Iron(II)} \quad Fe^{2+} + K^+ + Fe(CN)_6^{3-} \longrightarrow \underset{\text{Prussian blue}}{KFe(Fe(CN)_6)\downarrow}$$

some involve oxidation-reduction:

$$\text{Nitrate(V)} \qquad NO_3^- + 6H_2O + 8e^- \xrightarrow[\text{NaOH}]{\text{Al}} NH_3 + 9OH^-$$

and others involve complex formation:

$$\text{Cobalt(II)} \qquad \underset{\text{(pink)}}{Co(H_2O)_6^{2+}} \xrightarrow[\text{HCl}]{\text{conc.}} \underset{\text{(blue)}}{CoCl_4^{2-}}$$

Cation	*Result obtained from preliminary tests with* H_2S(12.2), NaOH *and* NH_4OH (12.3)	*Confirmatory test(s)* (on original solution, unless stated)
Aluminium Al^{3+}	No other cation has white ppt. soluble in excess NaOH but no ppt. with H_2S	The precipitate of $Al(OH)_3$ is very good at adsorbing dyes (e.g. alizarin) and becoming coloured
Barium Ba^{2+}	Only Ca^{2+} and Mg^{2+} also have white hydroxide ppts. insoluble in excess, but no ppt. with H_2S	Add dilute H_2SO_4: *thick* white ppt. of $BaSO_4$. To distinguish from Ca^{2+}, yellow ppt. of $BaCrO_4$ is insoluble in dilute acetic acid
Calcium Ca^{2+}	See Ba^{2+}	Add dilute H_2SO_4: *thin* white ppt. of $CaSO_4$. Yellow ppt. of $CaCrO_4$ is soluble in dilute acetic acid
Chromium Cr^{3+}	A pale green hydroxide ppt. soluble in excess NaOH is distinctive	Add dilute H_2O_2 to ppt. opposite and warm. Yellow CrO_4^{2-} formed. $2Cr(OH)_3 + 3H_2O_2 + 4OH^- \longrightarrow 2CrO_4^{2-} + 8H_2O$
Cobalt Co^{2+}	Hydroxide ppt. darkening in air like Mn^{2+}: but sulphide ppt. is black like Ni^{2+}	Add conc. HCl, and colour changes from pink to blue $Co(H_2O)_6^{2+} \longrightarrow CoCl_4^{2-}$
Copper Cu^{2+}	Deep blue solution with excess ammonia is characteristic	Add KI solution and observe white ppt. of CuI in brown solution of I_2 $Cu^{2+} + 2I^- \longrightarrow CuI + \frac{1}{2}I_2$
Iron Fe^{2+}	Green hydroxide ppt. turning brown in air is characteristic	Add solution of $K_3Fe(CN)_6$ (hexacyanoferrate(III)). Deep blue of Prussian blue, $KFe^{II}Fe^{III}(CN)_6$
Iron Fe^{3+}	Red-brown hydroxide ppt. is characteristic	Add solution of $K_4Fe(CN)_6$ (hexacyanoferrate(II)). Deep blue ppt. of Prussian blue, $KFe^{II}Fe^{III}(CN)_6$
Lead Pb^{2+}	The only black sulphide ppt. combined with a white hydroxide ppt. soluble in excess NaOH	Add KI solution. Yellow ppt. of PbI_2
Magnesium Mg^{2+}	See Ba^{2+}	Unlike Ba^{2+} and Ca^{2+}, no ppt. with dilute H_2SO_4
Manganese Mn^{2+}	Hydroxide ppt. darkening in air like Co^{2+}, but sulphide ppt. is characteristic buff colour	–

Cation	*Result obtained from preliminary tests with* H_2S(12.2), NaOH *and* NH_4OH (12.3)	*Confirmatory test(s)* (on original solution, unless stated)
Mercury Hg^{2+}	Red ppt. (of HgO) with NaOH is characteristic	Add $SnCl_2$ solution. White ppt. of HgCl, followed by greying as Hg formed
Nickel Ni^{+}	Green hydroxide ppt. soluble in excess NH_4OH to *light* blue solution is characteristic	Pink ppt. of Ni^{2+} in square planar complex with dimethylglyoxime in dilute ammonia solution
Potassium K^+	Only Na^+ and NH_4^+ also give no ppt. in these tests	Add $Na_3Co(NO_2)_6$ (hexanitrocobaltate(III)) solution. Yellow ppt. of K^+ salt
Silver Ag^+	Brown ppt. (of Ag_2O) with NaOH, soluble in NH_4OH, is characteristic	Add NaCl Solution, white ppt. of AgCl; or KI solution, pale yellow ppt. of AgI
Sodium Na^+	See K^+	No satisfactory test
Tin Sn^{2+}	The only brown sulphide ppt. combined with a white hydroxide ppt. soluble in excess NH_4OH	Add $HgCl_2$ solution. White ppt. of HgCl, greying as Hg formed $Sn^{2+} + 2Hg^{2+} \longrightarrow Sn^{4+} + 2Hg^+$: $Sn^{2+} + 2Hg^+ \longrightarrow Sn^{4+} + 2Hg$
Zinc Zn^{2+}	The only white hydroxide ppt. soluble in both NaOH and NH_4OH	

This table of confirmatory tests for ions applies only if single substances are present (or substances in which other ions present do not interfere with the test). For the separation and analysis of a multi-ion mixture you should see a work of reference.

Table 12.8 Confirmatory tests for cations.

Anion	*Result obtained from preliminary tests with acids* (12.5) *and* $AgNO_3$ (12.6)	*Additional confirmatory tests* (on solution unless stated)
Bromide Br^-	Only Cl^- and I^- also have silver ppts. insoluble in dilute HNO_3	Add Cl_2 in CCl_4 and shake. Brown colour of Br_2 in CCl_4 layer confirms Br^- $Cl_2 + 2Br^- \longrightarrow 2Cl^- + Br_2$
Carbonate CO_3^{2-}	Effervescence of colourless gas with dilute acid characteristic	Confirm presence of CO_2 with lime water
Chloride Cl^-	See Br^-	As for Br^-; but result of test is negative
Chromate CrO_4^-	Red ppt. of Ag_2CrO_4 is characteristic	Add H_2O_2 to acidified solution. Deep blue colour, transient and decomposing to O_2 and green Cr^{3+} solution, is characteristic
Fluoride F^-	Glass etching test with conc. H_2SO_4 is characteristic	–

Anion	*Result obtained from preliminary tests with acids* (12.5) *and* $AgNO_3$ (12.6)	*Additional confirmatory tests* (on solution unless stated)
Iodide I^-	See Br^-	As for Br^-: purple colour in CCl_4 layer is positive for I^- $Cl_2 + 2I^- \longrightarrow 2Cl^- + I_2$
Nitrate(III) NO_2^-	Effervescence of brown gas with dilute acid is characteristic	See redox tests (12.8) and below
Nitrate(V) NO_3^-	Largely negative. Compare with SO_4^{2-}	Warm with aluminium foil and dilute NaOH. Ammonia evolved confirms NO_2^- or NO_3^-
Phosphate(V) PO_4^{3-}	Bright yellow ppt. of Ag_3PO_4 characteristic	–
Sulphate(IV) SO_3^{2-}	Evolution of pungent SO_2 with dilute acid characteristic	–
Sulphate(VI) SO_4^{2-}	Largely negative. Compare with NO_3^-	White ppt. with $BaCl_2$ solution insoluble in dilute HCl $Ba^{2+} + SO_4^{2-} \longrightarrow BaSO_4$
Sulphide S^{2-}	Evolution of bad-egg smell of H_2S with dilute acid is characteristic	See redox tests (12.8) or sulphide precipitates (12.2)

Table 12.9 Confirmatory tests for anions.

QUESTIONS

1. In each case give a test which would distinguish between ions stated. Your test should give procedure, expected observations for each member of the pair, and an equation.

(a) Sn^{2+}(aq) and Zn^{2+}(aq)
(b) Fe^{3+}(aq) and Fe^{2+}(aq)
(c) NH_4^+ and K^+ as solid chlorides
(d) CO_3^- and SO_3^{2-} as solid sodium salts
(e) Ag^+(aq) and Pb^{2+}(aq)
(f) Co^{2+}(aq) and Ni^{2+}(aq)
(g) Br^- and NO_3^- as solid potassium salts
(h) Cl^-(aq) and Br^-(aq)
(i) Al^{3+}(aq) and Ca^{2+}(aq)
(j) NaOH(aq) and NH_3(aq)

2 (a) 'Cadmium yellow' is a recent development as a yellow artists' pigment. What might it be, and why is it superior to lead(II) chromate(VI) in a polluted atmosphere?

(b) Make an 'order-of-magnitude' calculation of the relative costs of
(i) using a spectrophotometer for analysis,
(ii) doing a manual analysis.

Allow for paying off the capital cost over ten years, interest charges during this time, and the labour costs involved in either method for ten analyses per working day. Do the relative advantages alter much if one analysis per day or one hundred analyses per day are performed?

(c) Is there any correlation between the solubility products of sulphides (this chapter) and the occurrence of metals as sulphide ores (chapter 6)? Would you expect this?

3 Explain
(a) why zinc sulphide is insoluble in acid but not in water, but copper(II) sulphide is insoluble in water and dilute acid;
(b) why aluminium hydroxide may be safely predicted to dissolve in sodium hydroxide;
(c) why sulphates(IV) give off sulphur dioxide when treated with dilute hydrochloric acid, but sulphates(VI) do not give off sulphur trioxide;
(d) why nitrate(III) solution readily oxidises acidified potassium iodide solution but nitrate(V) does not;
(e) why solutions of sodium and ammonium hydroxides do not re-dissolve the same hydroxide precipitates.

4 Identify the following and write equations for the reactions involved.
(a) A green solution gives a white precipitate with barium chloride solution in dilute hydrochloric acid: it yields a black sulphide precipitate in neutral but not acid solution: with dilute ammonia the faint green precipitate first produced redissolves easily to a blue solution.
(b) White crystals give off a brown gas which oxidises iodide to iodine, leaving a metallic residue: the crystals are readily soluble in water, giving a brown precipitate with sodium hydroxide and a red precipitate with potassium chromate.
(c) A white solid, insoluble in water, dissolves in dilute hydrochloric acid giving off a pungent choking gas which turns acidified potassium chromate(VI) from yellow to green: the solution so obtained gives a white precipitate dissolving in excess with both sodium hydroxide and dilute ammonia.
(d) A white solid, gives off purple fumes on treatment with concentrated sulphuric acid; itself dissolves in water and gives a white precipitate with sodium hydroxide insoluble in excess; the aqueous solution does not give a precipitate with dilute sulphuric acid.
(e) A reddish-brown solid, gives off white fumes with concentrated sulphuric acid, itself dissolves in water, giving a pale yellow precipitate with hydrogen sulphide in acid solution, but a black precipitate in neutral solution.
(f) A white solid, sublimes readily, and effervesces with dilute acid, the gas evolved reacting to turn lime water milky. Its aqueous solution shows no visible reaction with sodium hydroxide, but gives a white precipitate with silver nitrate soluble in dilute ammonia.
(g) A yellow solid, insoluble in water and cold dilute hydrochloric acid, soluble in dilute nitric acid without evolution of gas to give a colourless solution; this solution gives precipitates with solutions of hydroxides, chlorides, iodides and sulphides respectively white, white, yellow and brown. The original solid shows no oxidising or reducing properties.

Index

Electronic configurations of the elements (ground state)

Atomic number	Element	Principal quantum number										
		$n = 1$	2		3			4				5
		1s	2s	2p	3s	3p	3d	4s	4p	4d	4f	5s
1	Hydrogen	1										
2	Helium	2										
3	Lithium	2	1									
4	Beryllium	2	2									
5	Boron	2	2	1								
6	Carbon	2	2	2								
7	Nitrogen	2	2	3								
8	Oxygen	2	2	4								
9	Fluorine	2	2	5								
10	Neon	2	2	6								
11	Sodium	2	2	6	1							
12	Magnesium	2	2	6	2							
13	Aluminium	2	2	6	2	1						
14	Silicon	2	2	6	2	2						
15	Phosphorus	2	2	6	2	3						
16	Sulphur	2	2	6	2	4						
17	Chlorine	2	2	6	2	5						
18	Argon	2	2	6	2	6						
19	Potassium	2	2	6	2	6		1				
20	Calcium	2	2	6	2	6		2				
21	Scandium	2	2	6	2	6	1	2				
22	Titanium	2	2	6	2	6	2	2				
23	Vanadium	2	2	6	2	6	3	2				
24	Chromium	2	2	6	2	6	5	1				
25	Manganese	2	2	6	2	6	5	2				
26	Iron	2	2	6	2	6	6	2				
27	Cobalt	2	2	6	2	6	7	2				
28	Nickel	2	2	6	2	6	8	2				
29	Copper	2	2	6	2	6	10	1				
30	Zinc	2	2	6	2	6	10	2				
31	Gallium	2	2	6	2	6	10	2	1			
32	Germanium	2	2	6	2	6	10	2	2			
33	Arsenic	2	2	6	2	6	10	2	3			
34	Selenium	2	2	6	2	6	10	2	4			
35	Bromine	2	2	6	2	6	10	2	5			
36	Krypton	2	2	6	2	6	10	2	6			
37	Rubidium	2	2	6	2	6	10	2	6			1
38	Strontium	2	2	6	2	6	10	2	6			2
39	Yttrium	2	2	6	2	6	10	2	6	1		2
40	Zirconium	2	2	6	2	6	10	2	6	2		2
41	Niobium	2	2	6	2	6	10	2	6	4		1
42	Molybdenum	2	2	6	2	6	10	2	6	5		1
43	Technetium	2	2	6	2	6	10	2	6	6		1
44	Ruthenium	2	2	6	2	6	10	2	6	7		1
45	Rhodium	2	2	6	2	6	10	2	6	8		1
46	Palladium	2	2	6	2	6	10	2	6	10		
47	Silver	2	2	6	2	6	10	2	6	10		1
48	Cadmium	2	2	6	2	6	10	2	6	10		2

Atomic number	Element	*Principal quantum number*														
		n = *1*	*2*	*3*	*4*				*5*				*6*			*7*
					4s	*4p*	*4d*	*4f*	*5s*	*5p*	*5d*	*5f*	*6s*	*6p*	*6d*	
49	Indium	2	8	18	2	6	10		2	1						
50	Tin	2	8	18	2	6	10		2	2						
51	Antimony	2	8	18	2	6	10		2	3						
52	Tellurium	2	8	18	2	6	10		2	4						
53	Iodine	2	8	18	2	6	10		2	5						
54	Xenon	2	8	18	2	6	10		2	6						
55	Caesium	2	8	18	2	6	10		2	6			1			
56	Barium	2	8	18	2	6	10		2	6			2			
57	Lanthanum	2	8	18	2	6	10		2	6	1		2			
58	Cerium	2	8	18	2	6	10	2	2	6			2			
59	Praseodymium	2	8	18	2	6	10	3	2	6			2			
60	Neodymium	2	8	18	2	6	10	4	2	6			2			
61	Promethium	2	8	18	2	6	10	5	2	6			2			
62	Samarium	2	8	18	2	6	10	6	2	6			2			
63	Europium	2	8	18	2	6	10	7	2	6			2			
64	Gadolinium	2	8	18	2	6	10	7	2	6	1		2			
65	Terbium	2	8	18	2	6	10	9	2	6			2			
66	Dysprosium	2	8	18	2	6	10	10	2	6			2			
67	Holmium	2	8	18	2	6	10	11	2	6			2			
68	Erbium	2	8	18	2	6	10	12	2	6			2			
69	Thulium	2	8	18	2	6	10	13	2	6			2			
70	Ytterbium	2	8	18	2	6	10	14	2	6			2			
71	Lutetium	2	8	18	2	6	10	14	2	6	1		2			
72	Hafnium	2	8	18	2	6	10	14	2	6	2		2			
73	Tantalum	2	8	18	2	6	10	14	2	6	3		2			
74	Tungsten	2	8	18	2	6	10	14	2	6	4		2			
75	Rhenium	2	8	18	2	6	10	14	2	6	5		2			
76	Osmium	2	8	18	2	6	10	14	2	6	6		2			
77	Iridium	2	8	18	2	6	10	14	2	6	9					
78	Platinum	2	8	18	2	6	10	14	2	6	9		1			
79	Gold	2	8	18	2	6	10	14	2	6	10		1			
80	Mercury	2	8	18	2	6	10	14	2	6	10		2			
81	Thallium	2	8	18	2	6	10	14	2	6	10		2	1		
82	Lead	2	8	18	2	6	10	14	2	6	10		2	2		
83	Bismuth	2	8	18	2	6	10	14	2	6	10		2	3		
84	Polonium	2	8	18	2	6	10	14	2	6	10		2	4		
85	Astatine	2	8	18	2	6	10	14	2	6	10		2	5		
86	Radon	2	8	18	2	6	10	14	2	6	10		2	6		
87	Francium	2	8	18	2	6	10	14	2	6	10		2	6		1
88	Radium	2	8	18	2	6	10	14	2	6	10		2	6		2
89	Actinium	2	8	18	2	6	10	14	2	6	10		2	6	1	2
90	Thorium	2	8	18	2	6	10	14	2	6	10		2	6	2	2
91	Protoactinium	2	8	18	2	6	10	14	2	6	10	2	2	6	1	2
92	Uranium	2	8	18	2	6	10	14	2	6	10	3	2	6	1	2
93	Neptunium	2	8	18	2	6	10	14	2	6	10	4	2	6	1	2
94	Plutonium	2	8	18	2	6	10	14	2	6	10	6	2	6		2
95	Americium	2	8	18	2	6	10	14	2	6	10	7	2	6		2
96	Curium	2	8	18	2	6	10	14	2	6	10	7	2	6	1	2
97	Berkelium	2	8	18	2	6	10	14	2	6	10	8	2	6	1	2
98	Californium	2	8	18	2	6	10	14	2	6	10	10	2	6		2
99	Einsteinium	2	8	18	2	6	10	14	2	6	10	11	2	6		2
100	Fermium	2	8	18	2	6	10	14	2	6	10	12	2	6		2
101	Mendelevium	2	8	18	2	6	10	14	2	6	10	13	2	6		2
102	Nobelium	2	8	18	2	6	10	14	2	6	10	14	2	6		2
103	Lawrencium	2	8	18	2	6	10	14	2	6	10	14	2	6	1	2

PERIODIC TABLE

Key: Atomic number / Symbol / Name / Relative atomic mass

Period	Groups I and II (s Block)		d Block										Groups III (p Block)	IV	V	VI	VII	0
1	1 H Hydrogen 1·008																	2 He Helium 4·003
2	3 Li Lithium 6·941	4 Be Beryllium 9·012											5 B Boron 10·81	6 C Carbon 12·01	7 N Nitrogen 14·01	8 O Oxygen 16·00	9 F Fluorine 19·00	10 Ne Neon 20·18
3	11 Na Sodium 22·99	12 Mg Magnesium 24·31											13 Al Aluminium 26·98	14 Si Silicon 28·09	15 P Phosphorus 30·97	16 S Sulphur 32·06	17 Cl Chlorine 35·45	18 Ar Argon 39·95
4	19 K Potassium 39·10	20 Ca Calcium 40·08	21 Sc Scandium 44·96	22 Ti Titanium 47·90	23 V Vanadium 50·94	24 Cr Chromium 52·00	25 Mn Manganese 54·94	26 Fe Iron 55·85	27 Co Cobalt 58·93	28 Ni Nickel 58·70	29 Cu Copper 63·55	30 Zn Zinc 65·38	31 Ga Gallium 69·72	32 Ge Germanium 72·59	33 As Arsenic 74·92	34 Se Selenium 78·96	35 Br Bromine 79·90	36 Kr Krypton 83·80
5	37 Rb Rubidium 85·47	38 Sr Strontium 87·62	39 Y Yttrium 88·91	40 Zr Zirconium 91·22	41 Nb Niobium 92·91	42 Mo Molybdenum 95·94	43 Tc Technetium (97)	44 Ru Ruthenium 101·1	45 Rh Rhodium 102·9	46 Pd Palladium 106·4	47 Ag Silver 107·9	48 Cd Cadmium 112·4	49 In Indium 114·8	50 Sn Tin 118·7	51 Sb Antimony 121·8	52 Te Tellurium 127·6	53 I Iodine 126·9	54 Xe Xenon 131·3
6	55 Cs Caesium 132·9	56 Ba Barium 137·3	57 La Lanthanum 138·9	72 Hf Hafnium 178·5	73 Ta Tantalum 180·9	74 W Tungsten 183·9	75 Re Rhenium 186·2	76 Os Osmium 190·2	77 Ir Iridium 192·2	78 Pt Platinum 195·1	79 Au Gold 197·0	80 Hg Mercury 200·6	81 Tl Thallium 204·4	82 Pb Lead 207·2	83 Bi Bismuth 209·0	84 Po Polonium (209)	85 At Astatine (210)	86 Rn Radon (222)